Geology of the North-West European Continental Shelf

Volume 1

D Naylor & SN Mounteney

Graham Trotman Dudley Publishers Ltd.
London

First published in 1975 by

Graham Trotman Dudley Publishers Limited
20 Fouberts Place, Regent Street, London W1V 1HH

ISBN 0 86010 009 X

Printed and bound in Great Britain by
Redwood Burn Limited
Trowbridge & Esher

Contents

List of Illustrations

1 Introduction

The discovery of oil beneath the waters surrounding the British Isles is destined to have a greater impact on the commercial structure of this country than any other comparable event since the Industrial Revolution. As a result the exploration for, and the development, of new oil resources are of considerable interest to a number of people outside the relatively small group within the oil industry who actually plan and carry out these tasks.

From the banking community to the offshore supply industries there is a wide spectrum of personnel for whom a knowledge and understanding of the developments within the oil industry is of the upmost importance, as also is a broad understanding of why exploration activity is moving into new areas and the likelihood of success there. The basis of such knowledge must lie in an understanding of at least the outline geology of the continental shelves around the British Isles. For this purpose the following chapters have attempted to make this information readily available to the non-specialist.

Until recently our knowledge of these offshore areas had been fairly limited, but within the last decade there has been a very rapid increase in the amount of research activity throughout the world. As a result of the large volume of new factual data which has become available it is now difficult for anyone not actively working in these fields to fully grasp the new pattern of developments. This book explains some of these new findings and their effect on our understanding of the offshore region west of the British Isles.

The oil industry is now on the brink of a major phase of exploration within the offshore areas west of Britain and Ireland; a phase which will probably accelerate over the next few years, and eventually provide answers not only to some of the geological problems, but also to the question of whether offshore West Britain will repeat the phenomenal economic success of the North Sea oil province.

The aim of this book is to be intelligible to anyone with little or no geological background, although it has been assumed, whether for commercial reasons or purely for interest, that the reader is committed to an understanding of the first principles of the subject.

Thus whilst it is clearly impossible to incorporate a full introductory textbook of basic geology within the confines of this volume, a brief explanation of some of the major concepts has been included.

We wish to thank our colleagues in Exploration Consultants Ltd who have aided in the production of this book and upon whose ideas and work we have leaned heavily. In particular we are indebted to D R Whitbread, G Rees, E Dike and R M Pegrum. We must also express our deep gratitude to the following companies for the use of their seismic data in the production of several diagrams:- Western Geophysical Co, Compagnie Generale de Geophysique, Seiscom Ltd, Siebens Oil and Gas Ltd, Geophysical Service International Ltd.

The science of geology has undergone a considerable change within the last decade following the evolution of important new theories, in particular within the field of structural geology and crustal movement. These new developments are summarised in the first part of the book as they are pertinent to a full understanding of offshore West Britain. Basically the book provides information at three different levels. Firstly, the text by itself can be read with only a limited knowledge of geology, although in general the captions beneath the figures provide additional details. Secondly, for the reader with no geological background, the basic terms used are explained in a short Appendix and Glossary at the end of the book; and thirdly, for the really determined reader, there is a short (certainly not exhaustive) reading list at the end of each chapter.

The earth is a solid sphere of matter approximately 7,900 miles in diameter of which only the outermost surface can be observed through bare rock exposures at the surface or through data revealed by relatively shallow drilling surveys. The deepest oil wells have penetrated only five miles down into the earth's interior. Geological processes and volcanic eruptions can expose rocks that may have been created some 10 - 15 miles below the earth's surface, but apart from such scanty data there is no direct evidence for the structure of the earth's interior. However, indications of the internal composition have been made from geophysical studies such as the measurement of the gravitational and magnetic fields, and the passage of earthquake or shock waves through the sphere, in conjunction with other evidence such as the composition of meteorites.

On the basis of these observations the earth can be divided into three major zones. The outermost superficial layer is termed the *crust,* it averages a thickness of 15 - 25 miles beneath the continents and thins to less than 6 miles beneath the sea in the oceanic regions. The thick intermediate layer, termed the *mantle,* occupies approximately 83% of the earth's volume and is dividend into an upper and lower section roughly 500 miles below the surface. This is a very important unit as it is believed to be the source region of most of the earth's internal energy including those forces responsible for the movement of the overlying crustal plates described in Chapter 2. The third and innermost unit is known as the *core*. Like the mantle the core can similarly be divided into two sub-units, an outer fluid layer between 1,750 and 3,100 miles beneath the earth's surface and an inner solid layer.

From a study of the rocks which compose the crust it can be shown that they fall into three main categories: igneous rocks, sedimentary rocks, and metamorphic rocks. Igneous rocks, composed of tightly interlocking crystals of different minerals, are formed by the cooling and solidification of molten rock from the lower crust or upper mantle. This process is most dramatically shown during the eruption and rapid cooling of molten lava from volcanic vents or fissures in the earth's surface. In volcanically active areas molten rocks are also injected at depth into the crust without emerging at the surface, producing massive granitic bodies or sheet-like dykes and sills.

It is most likely that in the early part of the earth's long history the rocks at the surface were mainly of igneous type. But as the earth cooled and the agents of erosion, such as the wind and the rain became active, the existing igneous rocks became worn down and the eroded fragments and particles were carried down and deposited in lakes or rivers or out in the open sea. The accumulation and compaction of these erosional fragments led to the formation of *sedimentary rocks*. In general sedimentary rocks are the result of the accumulation and compaction of either clay, mud, silt, sand or shell fragments at the base of a body of water, or by chemical precipitation out of the water itself. Where the particles forming the sedimentary rock are relatively large, as in sandstones, compaction is relatively inefficient and open pore spaces of variable size and shape remain between the individual grains. These pore spaces may become infilled with a cementing material such as calcium or silica, or alternatively may remain open. Where open they are of considerable importance as they will be filled by fluids such as water or oil. Sedimentary rocks with open pore spaces provide the bulk of the world's oil and gas reservoir rocks.

Metamorphic rocks are the result of alteration of pre-existing igneous or sedimentary rocks by processes of heat, pressure or the action of hot solutions within the crust. Rocks originally formed at or near the earth's surface may become slowly buried to increasing depths within the crust, where they are acted upon by the dynamic forces of increased temperature and pressure. This can result in the rock being totally reformed (recrystallised) and is characterised by an absence of pore spaces. In general the older the rock the greater the opportunity that it may have undergone such processes.

From the facts given above, it follows that the sedimentary rocks are of prime interest to the petroleum geologist as these with few exceptions are the rocks which are likely to contain open pore spaces suitable for the accumulation of oil and gas, and also which provide the organic matter from which the hydrocarbons were originally derived.

Sedimentary rocks are deposited in sheets or layers known as *strata*, the study of which is termed *stratigraphy*. Since the strata are deposited one above the other, it follows that in an undisturbed sequence the deepest strata are the oldest, and the uppermost strata the youngest. Any large depression or sag in the earth's surface within which a thick sequence accumulates may be termed a *basin of deposition*. Normally sedimentary layers are deposited as a horizontal or very gently dipping surface, although locally, as in fan deposits along the fronts of mountain ranges, individual layers may be steeply sloping.

The study of the records of changing organic life, preserved within the rock strata as fossils, has led to the widespread acceptance by geologists of the theory of evolution. In general terms the theory demands a gradual change throughout time from primitive animals and plants to more complex organisms. It follows that in any thick sequence of sedimentary rocks the organic remains contained within them will show a gradual change. Each natural group of beds, or geological formation, ideally contains a distinctive suite of fossils which will be similar to those preserved in any other sequence or rocks deposited at the same time, and represents a measure of the level of evolution at that moment in geological time. The study of these fossil remains is *Palaeontology*. Palaeontological data allow the geologist to equate, or *correlate,* stratal sequences throughout the world. Although this method does not give the absolute age of the rock in years, it does allow for the age to be ascertained relative to another group of rocks. This method of correlation has been gradually built up by experience. It should be noted that some fossils evolve rapidly and are therefore of more use in correlation than other more slowly evolving types.

A sequence of strata developed in a local basin of deposition can thus be fitted into a much larger system correlatable on a world-wide basis. Within this world-wide system the relative time and the sequence of rock strata can be grouped into major units, with each sequence of rock layers being equivalent to a certain time span and being often characterised by a particular suite of fossils. The world-wide grouping of the stratal time units is shown in Figure 1. A major threefold division of fossiliferous rocks is recognised:

> Modern or Cainozoic Strata
> Middle or Mesozoic Strata
> Ancient or Palaeozoic Strata

These three divisions or *eras* are universally accepted and although they often coincide with distinct changes in rock type they are in fact primarily recognised by changes in their fossil content. Still older are the Proterozoic or Precambrian strata in which any records of life are both few and limited.

A further important concept in the study of sedimentary rocks is that of the *unconformity*. In any local succession of strata, certain beds can be absent, and as a result there is a break in the deposition and fossil record for that particular interval. This depositional gap may be caused by either the cessation of deposition during this time period or by the subsequent removal of the deposit prior to the deposition of the younger overlying beds. Important stratigraphic breaks are usually a reflection of a major event on the earth's surface which can be correlated over wide areas. When the dynamic forces at work in the crust give rise to crumpling, uplift and erosion of the strata, the overlying beds are deposited with *angular unconformity* on the tilted worn edges of the underlying strata (see Figure 34 in the Appendix).

Unconformities can represent a long break in the time period from which deposition is missing indicating widespread catastrophic events on the earth's surface, or only a short time break of a local nature, recognisable by the absence of certain fossils.

The term *hydrocarbons* refers to a wide range of compounds from methane gas to semi-solid tar and wax. Normally oil is found as a mixture of the liquid part of this range of compounds, which varies in density from one area to another. Often gas is contained in the oil under pressure. If the light oil fraction has been allowed to evaporate the heavy end of the fraction remains as a thick tar. On the other hand, gas can occur unassociated, without oil, as in the southern North Sea area. Oil often contains sulphur compounds and small amounts of complex non-hydrocarbons which can strongly affect the economics of refining the oil after it has been produced. Due to the relative immobility of the heavy fractions and the confined nature of the pore spaces they cannot be satisfactorily produced from deep reservoirs without specialist techniques and are mainly produced by fractioning of the crude oil during the refining process.

Hydrocarbons are derived from the decomposition of organic compounds. It is believed that much of this organic matter was incorporated as microscopic dead organisms in an oxygen-poor environment of mud deposition. Slow decomposition of the organic compounds followed as the mud was gradually buried under further layers of sediment. Throughout this time the action of anaerobic bacteria continued and as the mud was further compressed by increased burial chemical and physical processes

CHIEF DIVISIONS OF GEOLOGICAL TIME						
ERA	SERIES	SUBDIVISIONS	LOCAL NAMES	AGE IN MILLIONS OF YEARS	INTESITY OF MOUNTAIN BUILDING ACTIVITY	EARTH MOVEMENTS AND EUROPEAN PLATE MOVEMENTS
QUATE-RNARY	RECENT PLEISTOCENE			2		
TERTIARY (NEOGENE / PALEOGENE)	PLIOCENE	UPPER				
		MIDDLE				
		LOWER		7		
	MIOCENE	UPPER				
		MIDDLE				ALPINE MOVEMENTS
		LOWER		26		
	OLIGOCENE	UPPER				
		MIDDLE				
		LOWER		38		
	EOCENE	UPPER				
		MIDDLE				
		LOWER	FRIGG SAND	54		TERTIARY VOLCANISM IN N.W. SCOTLAND
	PALEOCENE	UPPER	FORTIES SAND			SEPERATION OF ROCKALL PLATEAU FROM GREENLAND
		LOWER	COD SAND DANIAN CHALK	63 - 64		LARAMIDE MOVEMENTS
MESOZOIC	CRETACEOUS	UPPER	CHALK TURONIAN LIMESTONE	100		EXTENSION OF SEA-FLOOR SPREADING INTO THE NORTH-ATLANTIC AND LABRADOR SEA
		LOWER				AUSTRIAN MOVEMENTS
			WEALDEN	136		LATE KIMMERIAN MOVEMENTS
	JURASSIC	UPPER	MALM			BEGINING OF SEPERATION OF SPAIN FROM NORTH AMERICA PLATE.
		MIDDLE	DOGGER			
		LOWER	LIAS	190 -195		EARLY KIMMERIAN MOVEMENTS
	TRIASSIC	UPPER	KEUPER			INITIATION OF LINEAR TROUGH SYSTEM
		MIDDLE	MUSCHELKALK			
		LOWER	BUNTER	225		
PALAEOZOIC	PERMIAN	UPPER	ZECHSTEIN			
		LOWER	ROTLIEGENDES	280		HERCYNIAN MOVEMENTS
	CARBONIFEROUS	UPPER				
		MIDDLE				
		LOWER		345		CALEDONIAN MOVEMENTS
	DEVONIAN (OLD RED SANDSTONE)	UPPER				
		MIDDLE				
		LOWER		410		
	SILURIAN	UPPER				
		MIDDLE				
		LOWER		440		
	ORDOVICIAN	UPPER				
		MIDDLE				
		LOWER		530		
	CAMBRIAN	UPPER				
		MIDDLE				
		LOWER		570		
PRE-CAMBRIAN				Approx. 4600		

Figure 1
Chief Divisions of Geological Time

slowly altered the organic matter into petroleum compounds. Although the first petroleum compounds probably begin to form after only a few feet of burial, the process continues to occur until some considerable depth is reached.

With an increase in burial of the sediments, the particles become more and more compacted and the pore space diminishes so that the water and oil are laterally and vertically squeezed out of the sediment. In some cases the oil then collects within an adjacent rock, such as sandstone, which in its compacted state retains adequate pore space, while in others it leaks to the surface and is lost to the atmosphere. A statum capable of holding producible fluids is known as a *reservoir rock* and is generally of sandstone or limestone composition. The relative amount of pore space between the grains in a sediment is called *porosity*. The ability of the trapped fluids to move within the reservoir rock is dependent upon the pore spaces being interconnected. This property can be measured and is expressed as *permeability*.

It is a common misconception where large oil or gas fields exist, that the hydrocarbons are thought to occupy large cave-like cavities or pools beneath the surface. It should be noted that the hydrocarbons exist only in the minute spaces between the grains in the same way that water saturates surface rocks. In the absence of hydrocarbons the pore spaces are filled with water.

If the oil is to be prevented from leaking to the surface and evaporating, certain conditions are required:-
i.   a *seal* or *caprock*, that is an impermeable barrier to prevent upward migration from the reservoir rock.
ii.   a *trap*, forming a closed or sealed structure which may result from several processes. Contours on the upper sealed surface of the reservoir are closed like the topographical contours on a hill. The simplest form of trap is a *dome* or *anticline* of strata. The oil or gas being lighter collects in the upper part of the dome whilst the lower part of the reservoir is water saturated. For a more complete explanation of the entrapment of oil see the Appendix Figure 35.

As the dynamic forces within the earth's crust continue to operate, causing faulting, folding and tilting of the strata, an accumulation of oil may migrate from one trap to another. It follows however, that the longer these forces continue to be active the greater the chance that hydrocarbon accumulations will escape to the earth's surface and be lost or be destroyed by the prevailing high temperatures. It must also be noted that the older a group of rocks the greater the chance that the rock porosity will have been destroyed. Statistics on the world's hydrocarbon reserves bear out the above, showing the 92% of the world's oil reserves and 75% of the gas reserves lie in relatively young rocks of Mesozoic and Tertiary age, while the ancient rocks of Palaeozoic and Precambrian age contain little hydrocarbon reserves.

The continental shelves which form the shallow water area (less than 600 ft depth) surrounding the British Isles, have wide areas covered by thick layers of Mesozoic and Tertiary strata which are prospective for hydrocarbons. As offshore exploration and drilling technology has advanced it is natural that the oil industry has begun to operate in these new, high-potential marine areas. Of prime interest initially were those areas of continental shelf which lay adjacent to already known onshore oil or gas discoveries, such as was the case of the North Sea.

Interest in the southern North Sea was first generated by the Shell/Esso discovery in 1959 of the Groningen Gas Field in northeastern Holland (probably the third largest gas field in the world with recoverable reserves estimated at 65 trillion cubic feet). Geologists were then led to consider the oil and gas potential of the sediments existing in the southern part of Britain, in Holland, Denmark and in Germany and those which stretch across the floor of the North Sea. At first the main target for exploration in the southern North Sea was the Permian sandstone reservoir productive in the Groningen field. However, more recently as exploration has provided new insights into the geology of the North Sea, the search has not only spread geographically but also to a number of other potential hydrocarbon bearing target horizons.

Oil has been discovered in large quantities in the northern North Sea region and over very recent years exploration interest has spread to the continental shelf lying to the west of Britain and Ireland. The generally unpromising appearance of the Precambrian and Palaeozoic metamorphosed rocks, which form much of Ireland and the western part of England and Scotland, largely detracted from an earlier interest in the adjacent offshore shelf as a potentially hydrocarbon bearing zone. Only gradually has it been realised that the small patches of younger Mesozoic and Tertiary sediments which locally overlie the ancient metamorphosed rocks were in fact the onshore margins of extensive offshore sedimentary basins stretching westward across the continental shelf.

Credit must largely go to the Universities for their part in initially establishing the geological outlines of these offshore basins over the past two decades by taking shallow seabed samples and detailed gravity and magnetic measurements across the continental shelf. The university teams have been able to establish the presence of Mesozoic and Tertiary rocks within this offshore area. From the gravity measurements obtained it has been

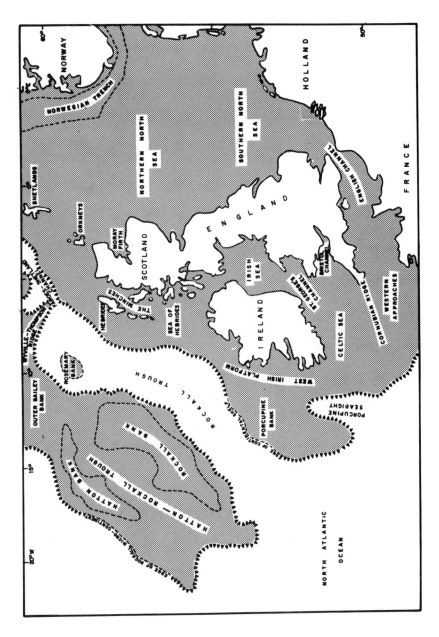

Figure 2
Morphology of the Continental Shelf of the British Isles.

possible to arrive at some estimate of the thickness of the young
sediments infilling these basins.

The procedural method an oil company adopts in exploring a new
area or thick basin-shaped accumulation of young sediments,
whether onshore or offshore, follows a four phase pattern. In the
first phase an examination of the exposed rocks at the earth's surface is carried out to establish their relative age and type. As

15

mentioned earlier the older the rock the less chance there is of finding oil. Since reservoir rocks may include ancient limestone reefs, sand deltas, sandbars and their original depositional position is controlled by the geography of that particular time period (palaeogeography), then a detailed study of rocks of the same age within the area can lead to an understanding of the palaeogeography and hence the possible position of reservoir rocks within the target zone. Similarly a study of the geological structure of the area, ie the way in which the rocks have been folded and buckled since compaction, may indicate a possible method of hydrocarbon entrapment.

Initially the obvious place to begin looking for such information is in land areas where the rocks can be seen at the ground surface. However, in many topographically low areas these older rocks of interest have been buried beneath a cover of younger sediments. Mountainous areas usually have good exposures of older rocks, and for this reason, field parties of oil company geologists go into remote mountain areas to make a detailed examination of the rocks as the first phase in exploration for oil.

In the offices and laboratories the geologists make a detailed study of samples collected. Fossils are studied in order to work out the relative ages of the strata. The geologist then tries to co-ordinate this outcrop information with *subsurface information* obtained by drilling wells (small chips and cores of rocks are taken whilst drilling). Other drilling records, mentioned below, are also of use. The fossils are used to correlate from one well to another. The end product of this study is a three dimensional model of the strata, which may occupy a large basin-shaped depression, tens or hundreds of miles in diameter. From this model the more promising areas for further exploration can be selected.

In the next phase of exploration, effort is concentrated on the search for potential oil traps. These may be buried at great depth. Depths between 5,000 — 20,000 feet are quite normal figures to be considered for exploration by drilling. Unless there is a great density of wells within the area, the geologist can at the best only suggest the type of trap to be anticipated and select the most favourable area for hydrocarbon accumulation.

The next tool in the search is the use of *seismic techniques*. When dynamite is exploded in a hole at the earth's surface part of the energy travels down into the earth. Some of this energy is then reflected back to the surface from the strata at depth and can be recorded (see Appendix Figure 36). Some rock layers, particularly at the junction of two very different rock types, give good reflec-tions. The energy returning to the surface can be initially recorded on tape and later, after computer processing, printed on paper. The end result is a *seismic record* which plots energy returned against time. If a number of records are lined up, then potential trapping structures such as faults and anticlines can be seen. Figure 16 B shows an example of a seismic record.

A number of changes in seismic techniques have occurred over the past ten years. The major advance has been in the application of the digital computer to processing the initial results and as a result, the use of multiple shots (explosions) and different recording techniques are now possible and are used extensively. Many processes which previously had to be laboriously carried out by hand, can now be effected by replaying the digital tapes. Extraneous energy (termed 'noise') on the records can be removed with the use of filters and the result is much higher clarity in the seismic records. Exactly the same techniques are used in offshore exploration except that the charge is exploded in the water and the geophones are towed on a cable behind a ship (Appendix Figure 36). Alternatives have also been found for conventional dynamite explosives, and extensive use is now being made not only of impact energy sources such as explosive airguns and gas guns but also implosive sources.

Taking the seismic records, the geologist and geophysicist attempt an interpretation of the stratigraphy and structure of the area. They may come to the conclusion that one or more potential traps are present. Refinement of these ideas by further detailed seismic and geological work may follow. Ultimately, however, the next stage is to test the hypothesis by drilling an exploratory hole.

The decision to drill a 'wildcat' (ie exploratory) well is a complex managerial decision. The cost of drilling, the size of the trap, the amount of oil which might be expected if successful, the distance to market and many other factors enter into the judgement. Once the decision has been taken, however, the operation becomes an engineering problem.

An oil well is normally drilled by turning a rotary table attached to a pipe, bearing a tungsten carbide or diamond bit. Additional pieces of pipe are coupled on at the top as the well deepens. When the bit has to be changed the whole process is reversed, each length of pipe is pulled out, uncoupled and stacked on the inside of the derrick. The bit changing operation may take a day if the well is at great depth. The deeper the intended hole is to be, the larger the rig and engines required for pulling and stacking the huge lengths of pipe, and for applying weight on the drill bit. A diagram showing the equipment used in drilling is given in the Appendix (Figure 37). The drilling bit cuts small pieces of rock at the base of the hole, and these are flushed out by a stream of mud which is continuously pumped down the inside of the pipe and back up the annulus between the pipe and the side of the hole to the surface. There the rock chips are collected and examined by the geologist and progress of the well towards its objective assessed. The mud also acts as a lubricant for the bit and by the addition of barytes to increase its weight prevents oil or gas contained under pressure at depth from suddenly blowing out when penetrated by the bit. At the well head there are blowout preventers which operate if excessive pressures build up.

One drawback of drilling with heavy mud under pressure is that it is possible to drill through an unexpected oil zone without its presence being recognised as mud under pressure is forced laterally into the pores of the rock and prevents any oil escaping into the mud stream. In this way oil and gas fields have been missed in the past only to be discovered by subsequent drilling. Ideally the well is drilled close to hydrostatic balance so that when a hydrocarbon bearing sequence is penetrated sufficient traces enter the mud stream which can be detected either visually, as oil stained rock chips, or by gas detecting instruments which constantly monitor the mud returns at the surface.

When the hole has been drilled to total depth (TD) various electromechanical 'logging' tools can be lowered down it. One such tool attempts to collect a sample of the reservoir fluid, others measure differences in resistivity, the speed of passage of sound waves, radioactivity and other features. Each method provides the geologist with information regarding the rock type and its porosity and permeability at a particular level.

If the drillhole is approaching the predicted depth of an important reservoir stratum, the geologist may ask for a rock core. To obtain this the drill pipe is pulled out and a core barrel attached. After returning to the hole the drill pipe is turned with the core barrel sitting at the base of the hole and a complete rock core several inches in diameter, and up to 60 feet in length, may be cut and is then raised to the surface. The expense of this operation may be great due to the extra time involved. Every 60 feet or less the entire drill has to be removed to recover the core whereas a conventional bit may drill many hundreds of feet between bit changes. Therefore the minimum number of cores is cut in any one hole.

When a productive horizon has been penetrated, a controlled test is run whereby the oil or gas is allowed to flow to the surface and escape through a small choke. The size of the reservoir and production potential of the well can then be ascertained from the results of the test.

As can be seen in Figure 37 there are three basic types of offshore drilling rig in use around the British Isles:

i   **Floating Rigs:** Usually purpose built or converted ships which are anchored over the location. The main advantages are the extreme manoeuvrability (in that the rig can move to the next site under its own power) and the great depths of water in which it is possible to drill.

ii   **Semi-Submersible Rigs:** These are specially constructed platforms standing on a pontoon substructure which can be partially filled with water. The whole structure floats but has greater stability than a ship, as the supporting pontoons can be positioned below the normal wave base where movement of the water is considerably less. The rig is held in position by multiple anchors. A disadvantage is that in most cases these rigs must be towed to the next site and are more vulnerable to instability during this time. Designs are very variable and some semi-submersibles are capable of self propulsion.

iii   **Jack-up Rigs:** These have extendable legs which can be withdrawn so that the rig can be floated to the next location. On location the legs are then extended until they rest on the sea-floor raising the drilling platform clear of the water on hydraulic jacks. Such rigs provide a stable drilling base but limitations imposed by the length and weight of the legs, and their stability during towing, restrict them to the shallower water areas, generally less than 350 feet.

Special techniques are required at sea to complete wells for oil or gas production. Usually a platform is constructed from which the follow-up or development wells are drilled in order to produce the oil or gas. A permanent platform is usually constructed of steel or concrete, and is left in position for the life of the field, and the oil or gas piped from there to the shore. Exploration is gradually moving into deeper waters which are beyond the depths of conventional production platforms, and currently sea-bed completion techniques are being developed which will allow production equipment to be installed on the sea-floor, from which oil or gas will flow to shore or submarine storage tanks. In general, drilling costs are higher offshore, although remote onshore locations are also expensive. The total overall cost of a North Sea well may be in the order of £2million.

In order to fully understand the activities of the oil industry off the British shores, it is necessary to look briefly at the history of the industry. A factor to be borne in mind is that during the nineteenth and early twentieth centuries the principle and basic understanding of the science of geology had already been broadly developed. It was thus possible within a short period of the fortuitous inception of the industry, for the exploration for hydrocarbons to be given a broad scientific basis.

If we define the oil industry as the drilling of boreholes with the intention of finding, producing and marketing oil, then the industry is more than one century old. The first well to conform to the definition was drilled in Titusville, Pennsylvania, in 1859. The use of naturally occuring oil from surface seepage, however, goes back to at least 3,000 BC in Mesopotamia. Oil seepages on the ground surface occur in many parts of the world, and the location of major seeps were known to many ancient civilisations. From the earliest times oil was credited with great beneficial and healing properties.

The use of tar from oil seeps (produced when the lighter fractions of the oil have evaporated) for waterproofing boats was also understood from an early age. Important seeps such as the Pitch Lake in Trinidad, have been used by sailors for this purpose until comparatively recent times. Many of the uses to which oil was put by early civilisations were not rediscovered until the 19th Century. Distillation techniques were developed in the Middle East in pre-Christian times and the light fraction used as an illuminant.

The early history of the modern oil industry took place in the United States. There are several reasons for this. There were vast oil reserves on the North American continent and some of the oil lay at relatively shallow depths, or appeared in surface seepages. Increasing industrialisation, the demand for better lighting, developing drilling techniques and easily accessible oil were all present together at the correct moment in mid-19th Century USA. Seepages were known from early settler days in many parts of North America. Oil was also encountered when drilling salt wells. Salt was a valuable preservative for food and the oil which spoiled the brine was a great nuisance to drillers of salt wells, particularly in Virginia. However, the techniques developed for the drilling of salt wells were to be invaluable in oil exploration.

With the growing belief in the medicinal properties of oil and its use for lubrication, oil from seeps was bottled for sale. Demand for better lamp fuels (previously animal and plant oils had been used) had given rise to a flourishing industry in distilling oil from coal. The small amounts of natural oil available from surface seeps were not enough to threaten this industry, although some trenching had been done in the area of seeps to increase the flow.

During the 1850s a group of people came to the conclusion that it would be possible to use salt well drilling techniques to get oil out of the ground (there were only vague theories at that time as to the true mode of its occurrence beneath the surface). The project was placed under the control of Col E L Drake, and after many set-backs, oil was recovered at shallow depth near Titusville, Pennsylvania in 1859. The amounts were small, but the method had been demonstrated.

Drake was followed by many others and within a short time the exploration for oil spread rapidly outwards from Titusville. The factors controlling entrapment of oil were recognised and the expansion of the oil industry began. The demands for illumination, followed by the internal combustion engine, petrochemicals and numerous other uses to which hydrocarbons have been put, have insured the rapid growth and expansion of the industry.

Many industrial countries have developed to a point at which they are over-dependent upon oil. The finite nature of the world reserves is now widely recognised and in future the search for hydrocarbons will be forced to extend into far more physically demanding environments and where drilling depths are greater. Such factors make great demands on new technology. This development can be seen in action around the shores of the British Isles at the present time. New generations of drilling equipment are enabling the boring of new holes in increasing depths of water. Already offshore production of oil is common in up to 300 feet of water supported by new diving and completion techniques. Production from water more than 400 feet deep is planned for the next few years and by the late 1970's this depth will have been increased to 600 feet or beyond.

To summarise it can be said the oil industry is poised to begin a major exploration effort on the continental shelves west of Britain. There are many interwoven factors which have brought about this situation, amongst which are:

i   Increased knowledge of the structure and geology of continental shelf areas.

ii   Improved and relatively inexpensive ship-borne seismic techniques which allow a picture of the subsurface to be constructed.

iii   Improved drilling equipment and technology which allow operations in increasing water depths in the stormy seas around Britain.

iv   Development of improved well completion techniques and production platforms which will allow production from fields on the deeper parts of the continental shelves within the next few years.

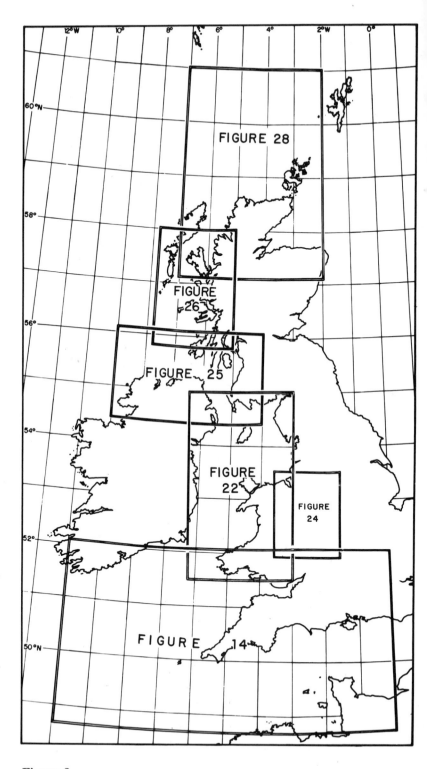

Figure 3
Map showing location of other Text Figures.

v   The increasing urgent need to find more oil for the industrial nations.

It must be realised that offshore West Britain and Ireland have so far received only preliminary exploration attention. Seismic surveys and sea-floor sampling have indicated the geological and geophysical potential of the region, but so far only a handful of deep wells have tested the prospects. The next five years will determine whether the geological promise is justified in terms of large and commercial volumes of hydrocarbons.

Selected Reading

HOLMES, A   1965   Principles of Physical Geology.
                                 Nelson 1,288 pp.

NAYLOR, D   1972   The hydrocarbon potential of Western Britain and Ireland. North Sea Conferences 1 & 2. I P C Industrial Press 130 — 137.

THE OPEN   1973   The Earth's Physical Resources 2. Energy Resources. 39 — 77.
UNIVERSITY

WHITBREAD, D R W   1972   The hydrocarbon potential of Western Britain and Ireland. North Sea Conferences 1 & 2. I P C Industrial Press. 81 — 84.

2 Plate Tectonics North-West European Margin

Sea Floor Spreading and Plate Tectonics

The concept of *plate tectonics* or *sea-floor spreading* is based on an understanding that the entire surface of the earth is composed of a number of internally rigid, but relatively thin (60—100 miles thick) interlocking plates continuously in motion relative to one another. The rate of motion is very slow, but can be detected and measured over a period of years using very sensitive instruments.

The plates themselves can be subdivided into an *upper* and *lower unit*. The *upper unit* is composed of either thick (15—25 miles) low density continental crust, or thinner (5 miles thick), more dense *oceanic crust*. Continental crust characterises the land regions of the world and the adjacent zone of shallow-water marine shelf, while oceanic crust characterises the deeper water zone which floors the major oceans and seas. The underlying *lower unit* extends beneath oceanic and continental crust alike, and is composed of the relatively brittle uppermost section of the high density *mantle*, some 55—75 miles in thickness. These relatively rigid plates *float* on the more mobile underlying mantle layers which constitute a major portion of the earth's interior.

It can be noted from Figure 4 that the individual plates show a marked variation in size and shape, and that at the present-day, the earth's surface is occupied by seven major plates and a number of smaller plates. Although it is now generally accepted that both the earth's volume and surface area have remained roughly constant throughout geological time, geophysical experiments over the last few years have established fairly conclusively that new crustal material is being generated along the central axes of many of the large oceans today, and being consumed along other parts of their margins. In order to maintain a constant surface area, the total volume of new crustal material created must equal that destroyed by consumption, and it is the maintenance of this balance that leads to the continuous motion of these plates across the surface of the earth. Further, as the motion of all the plates is to some extent interdependent, any change in the motion of one plate must be accompanied by an equivalent motion elsewhere.

Because of both the irregularity in shape and rigid nature of the plates, it is the plate margins that are of most interest, since it is here, that as a result of differential movement between adjacent plates, the main tectonic activity of the crust (eg earthquake zones, volcanic belts and mountain ranges) is concentrated.

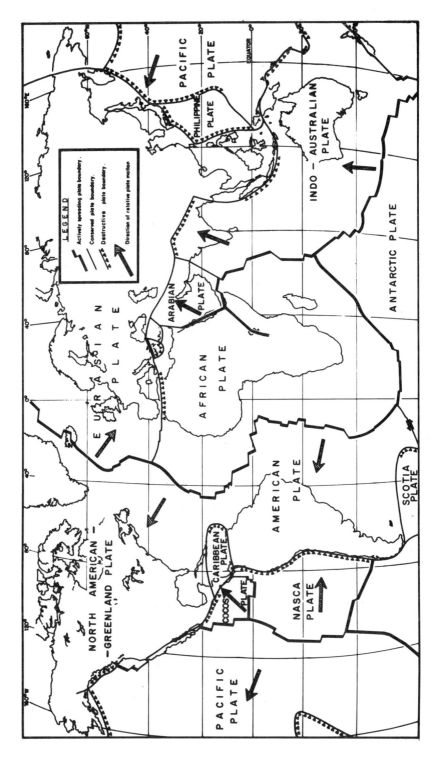

Figure 4
Major Plates of the World.

25

Three characteristic types of plate boundary are recognised:

1   A *constructive margin*, developed where two plates are
     moving directly away from each other, gives rise to
     expanding seas or oceans, or where it bisects a land-
     mass to a linear rift valley. Across the ocean floor the
     constructive plate boundary is marked by a linear
     chain of submarine mountains, termed the *mid-oceanic
     ridge*. New oceanic crust is continuously being
     generated along the length of the ridge during the pro-
     cess of plate separation, and when formed becomes
     effectively welded to the separating edges as part of the
     rigid plate, leaving no physical gap between the plates
     (see Figure 5). This new material produced at a con-
     structive margin can be regarded as forcing the two
     plates apart. In the early stages of rupture of a
     continental mass, separation of the two broken
     fragments is slow, but with the appearance of a
     juvenile ocean, and the initial generation of oceanic
     crust, the rates of separation speed up to reach values of
     up to a few centimetres a year. The most obvious
     consequences of this process is to produce a
     symmetrical sea-floor on either side of the ridge crest.

2   A *destructive margin*, developed where two plates are
     moving directly towards each other, marks a line where
     oceanic crust is being destroyed. Such a line generally
     occurs along the margins of continental masses where
     its surface expression is a system of oceanic trenches
     and island arcs. Where either one or both plates are com-
     posed of oceanic crust, the leading edge of one plate is
     forced beneath the other at an angle of about 45° (see
     Figure 6) becoming molten with depth and eventually
     being reabsorbed into the earth's mantle. However,
     although the huge blocks of continental crust behave
     as passive *passengers* on the plates, their low density
     composition renders them incapable of being forced
     beneath the edge of another plate, and this itself places
     significant restraints upon plate motion. Where two
     advancing plates both have an upper section composed
     of continental crust, then the relative buoyancy of this
     low density continental crust with respect to the under-
     lying mantle prevents the destruction of either plate
     boundary by plunging beneath the other, and direct
     collision occurs. This results in a gradual cessation of
     motion as the continental crust of both plates becomes
     welded together to form a single plate. The line of such
     a collision boundary is marked by a belt of strongly
     buckled and thickened continental crust taking the
     form of a mountain chain.

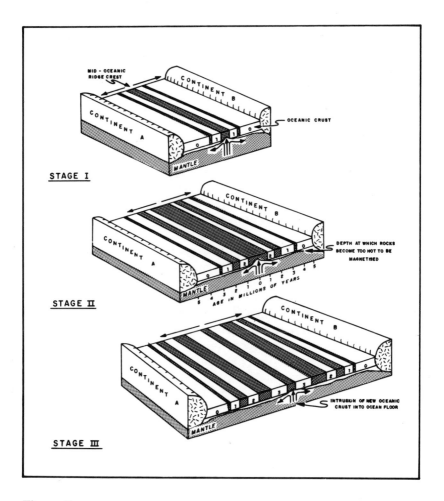

Figure 5
Diagrammatic representation of Plate Growth.

A diagrammatic representation of three stages in the growth of a new ocean floor showing the characteristic development of a pattern of symmetrical magnetic stripes about the central active spreading ridge. As molten material from the earths mantle is forced to the surface along the line of the mid-oceanic ridge it cools and is magnetised in the direction of the earth's field at that time, as shown by the pair of white stripes marked 'O'. Following a reversal in the direction of the magnetic field, all subsequent additions of new oceanic crust during that period (shown as the black band between 'O' and '1') are extruded along the central axis of the mid-oceanic ridge and cool to be magnetised in the direction of the new field, thereby separating and gradually pushing apart the two halves of the initially formed white band. As a result of a number of further magnetic reversals a pattern of both parallel and symmetrical magnetic stripes is built up about the central spreading ridge, with the width of the individual bands representing the time period between each reversal.

3   A *conserved margin* occurs where two adjacent plate margins are slipping past each other. As the relative motion between both plates is parallel, crustal material is neither created nor destroyed.

A point at which three plate boundaries meet is termed a *triple plate junction*.

Transform faults, perpendicular to the plate boundary and parallel to the main direction of movement, develop on the sea-floor to accommodate slight discrepancies in the spreading rate where the constructive plate boundaries are long and sinuous. Transform faults grow in length as the floor of the ocean expands.

Throughout geological time the direction of the earth's magnetic field, generated by its metallic core, has intermittently reversed polarity from South-North to North-South. Molten rocks rising from the mantle at a constructive plate boundary cool and solidify to form a new crust which is magnetised in the direction of the prevailing polarity. Once cooled, the rocks retain their original polarity, leading to alternating belts of oceanic crust magnetised in opposite directions which record the magnetic reversals. These magnetic *stripes* provide geologists and geophysicists with a quantative key to the rates at which two plates have separated. Not only does each polarity reversal exhibit a unique magnetic pattern but also by correlation with geological factors elsewhere, the time span during which any one particular magnetic polarity lasted, can also be fairly accurately calculated. As a result, the earlier position of two plates, now separated by an expanse of oceanic crust, can be reconstructed for any particular period simply by fitting together the pair of stripes for that date and ignoring the new crustal material produced since that time.

The ultimate causes which give rise to the development of new plate boundaries across previously stable and rigid continental masses are, as yet, only poorly understood. It appears, however, that the development of such a boundary does not always proceed to the actual point of continental rupture, which culminates in the growth of a new ocean. Throughout the world a number of failed attempts at plate separation can be seen crossing now stable blocks of continental crust. All tensional activity appears to have ceased along these lines and they are now frequently referred to as *abortive spreading aces*.

The preliminary stages in the separation of two continental plates and subsequent sequence of events leading up to the generation of an embryonic and expanding ocean, as depicted in Figure 8 produce a pattern of geological factors of considerable importance to the accumulation of potential oil and gas bearing sediments.

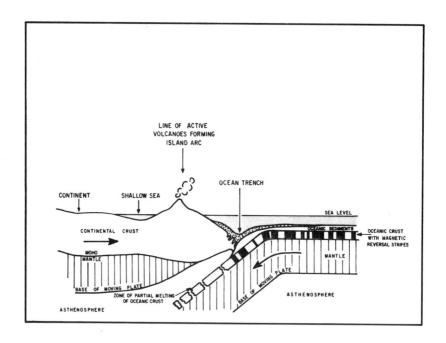

Figure 6
Destructive contact between two plates.

A destructive *boundary formed where two plates of different crustal composition are moving directly towards each other. The denser oceanic material of the plate on the right is being thrust beneath the lighter continental crust of the plate on the left. At a certain point the heat retained within the lower part of the earths crust (Asthenosphere) causes the descending plate to slowly loose its rigidity, melt and become partially reabsorbed into the asthenosphere. A certain amount of the slightly lighter molten material rises directly to the surface to be expelled through a chain of volcanic vents. Note the development of a characteristic pattern of deep linear trenches backed on the landward side by a chain of island arcs and volcanoes similar to that seen on the Japanese-Pacific collision boundary. Occasionally frictional resistance set up between the upper surface of the descending plate and the lower surface of the underthrust plate gives rise to a somewhat* jerky *pattern of downward movement. These* jerks or shock waves *are reflected at the surface as deep or shallow centred earthquakes, depending upon their place of origin along the descending plate boundary, and account for a continuous line of active earthquake epicentres along the entire* destructive *plate boundary.*

Initially tensional stresses distend the continental crust along the line of the developing plate boundary, creating a series of sub-parallel fractures into which molten material from the underlying mantle is injected. The molten material infills these fractures, and for the major part cools and solidifies in situ, to produce

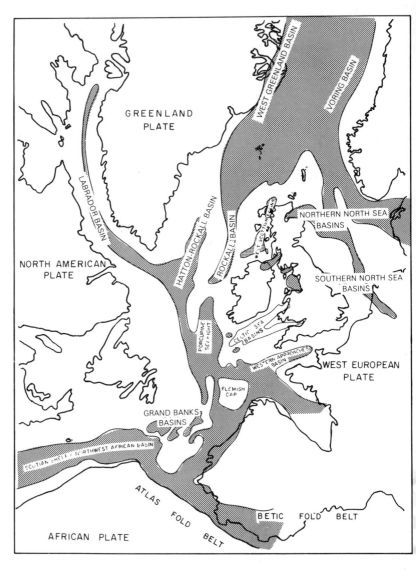

Figure 7
Reconstruction of North Atlantic plates prior to separation.

Throughout Permo-Triassic and early Jurassic times, prior to the initiation of ocean-floor spreading, the North Atlantic continents formed a single continental plate, in which the present day North Atlantic Ocean including the Bay of Biscay did not exist. The above reconstruction is based on fitting together the continental shelves bordering North America, Northeast Canada and Greenland into those bordering Northwest Europe and North Africa. The elongate shaded areas show the extent of the shallow shelf seas which existed in early Jurassic times. These seas follow the lines of a series of downfaulted grabenal troughs which developed in response to the pre-rupture phase of tensional stress between the North Atlantic plates.

a pattern of cross-cutting *dykes* within the upper crustal sediments; however, some escapes to the earth's surface and is expelled as molten material through a series of volcanic vents. Following further separation, these fractures show a marked increase in intensity as the system develops both laterally and longitudinally. This eventually results in partial collapse of the continental crust and the development of a chain of assymmetrical basins, such as those present in the continental shelf off the west coast of Britain, and finally, following major collapse of the central portion of the crust, to a deep, asymmetrical trough, many hundreds of miles in length.

Throughout these latter stages of crustal distension, the floor of the newly created central trough becomes the site of large lakes, and the accumulation of lacustrine and continentally derived sediments and volcanic debris from the flanking margins, or where sufficiently depressed to be invaded by the sea, is partially infilled by sediments deposited in a shallow marine environment.

Finally after considerable distension of the continental crust, the central trough is torn apart, and the newly created gap is invaded by molten oceanic crust which cools to produce a pair of new oceanic margins along adjacent edges of the separating plates. At the same time, the thick continental and volcanic sediments previously infilling the central trough are also torn into two sections and preserved along the separating continental margins of the growing ocean.

The present day Gulf of Suez-Red Sea-Gulf of Aden system illustrates the pattern of embryonic and juvenile ocean expansion described above. Here at the northern, younger end, the Gulf of Suez is barely 30 miles wide, and forms a fault-bounded trough underlain by thinned continental crust, and infilled with approximately 12,000 feet of young Tertiary and Recent sediments. Southwards, plate separation is more advanced and the Red Sea forms a broader trough with a thick marginal sedimentary sequence overlying a faulted and distended floor formed of partially ruptured continental crust with a central zone of expanding oceanic crust.

A study of ancient and present-day patterns of plate margins throughout the world indicate that the process of continental rupture was, and still is, frequently complicated by the development of a *triple plate junction*, with crustal extension attempting to occur simultaneously along three converging spreading axes. Such a triple spreading system is basically unstable and frequently results in active separation developing along only two of the separating arms, with the third, failed arm, becoming a dormant feature. The failed arm can be the site of a deep grabenal basin in which accumulated sediments may reach a thickness of

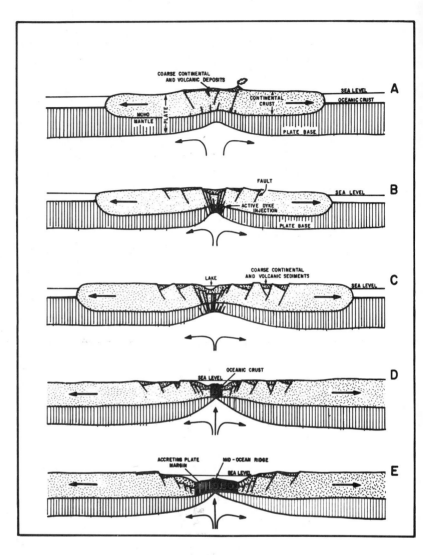

Figure 8
Diagram showing the initial stages of plate separation.

Schematic sections to illustrate the sequence of events in the rupturing of a continent and creation of a new ocean. Although the causes of both initial plate formation and subsequent plate motion are far from clearly understood, due to a general lack of knowledge of the physical properties and behaviour of the interior of the earth, it is thought that:-

A)   The development of a new pattern of mantle upwelling and lateral movement within the mobile earth's interior gives rise to a system of strong tensional forces in the relatively rigid continental crust above. These stresses cause tensional faults and fractures to develop in association with local downtilting of the crust, and are occasionally accompanied by volcanic activity as the molten material from below is forced to the surface along one such fracture.

30,000 feet. Such basins are of importance in the search for oil and gas reserves, since here the conditions of deposition approach an optimum for the generation, maturation, migration and accumulation of hydrocarbons.

2.2	**Major Rift Systems on the North-West European Shelf**

Throughout Europe, the North Sea, and over the offshore margin west of Britain, recent geological and geophysical data has shown a well defined pattern of deep, linear, sediment filled troughs, up to 30 miles wide and 200 miles long, separated by uplifted fault-bounded platforms of continental crust (see Figure 9). This pattern may represent a number of wholly or partially failed spreading systems.

The North European-North Sea rift system began to develop during the Late Palaeozoic, continuing through the Mesozoic period as the Asian-European-American megacontinent began to break up, and only subsequently becoming dormant in Late Mesozoic times as the locus of active crustal separation shifted westwards to follow the line now occupied by the North Atlantic Ocean and Labrador Sea.

Within Western Europe, three basic trough systems are apparent (see Figure 9).

---

continued

B-C) As the tensional stresses increase, the crust and underlying upper mantle forming the plate show considerable thinning with a marked increase in the number of both open and intruded fractures (dykes) and faults. Coarse, continentally derived sandstones and conglomerates, along with any local volcanic debris, infill these downfaulted depressions. The central depression often becomes so extensive as to be infilled by a large lake, eg. the East African Rift Valley, and eventually sufficiently low to be invaded by the sea, eg. the Red Sea.
D-E) Finally the plate ruptures and the two parts of the original continental mass suddenly become two separate plates moving directly away from each other. All tensional stresses are relieved and little further fracturing or faulting takes place in the vicinity of the newly created boundary. The growing gap between the two plates becomes simultaneously infilled with denser oceanic crust from the mantle below, and both parts of the original sediment-filled central basin become preserved along the edges of the new continental margins on either side of the ocean.

1   A North Sea System of interlinked troughs extending
     from the edge of the continental shelf west of Norway,
     through the North Sea and into southern Europe, via
     the Rhine graben system.

2   A Skagerrak System consisting of the Oslo rift and its
     continuation into the Skagerrak; the Danish-Polish
     Trough and the West Norway Trough.

3   A West Britain System extending from the West Shet-
     lands area through the Rockall Trough and the Rockall-
     Hatton Bank areas, and including the Porcupine Sea-
     bight and the Celtic Sea and Western Approaches
     troughs.

The Central North Sea trough system is almost 750 miles long
and constitutes a major structural feature on the geological map
of Europe shown in Figure 9.

A linear fault-bounded basin is present in the Celtic Sea south of
Ireland and north of the Cornubian Platform. The eastern end of
this trough divides, sending one arm into the Cardigan Bay area
of the Irish Sea, and one arm into the Bristol Channel between the
Pembroke Ridge and the Cornubian Platform. The entire trough
contains a thick Mesozoic section, unconformably overlain by
Upper Cretaceous and Tertiary sediments. A similar, but wider
fault-bounded Mesozoic trough occurs beneath the English
Channel, extending westwards to the continental margin as the
Western Approaches trough. Diagrammatic cross-sections
across all these linear basins are given in the relevant chapters
later in the book.

To date much of the commercial oil and gas found in the North-
West European shelf has been associated with the thick sediments infilling these fault-bounded troughs in the North Sea
area. A typically rifted origin, common to all these troughs has
led to a basic similarity in basin architecture and sedimentary
development, and may therefore, point to favourable oil exploration conditions in other troughs and platform flanks west of the
British Isles.

A striking worldwide feature of these rifted graben systems is the
repeated occurrence of triple plate junctions along their length
where three potential spreading axes converge on one point. The
location of possible triple plate junctions around the British Isles
are shown in circled numbers on Figure 9 Centres 1,2 and 5
(Figure 9),   located in the North Sea, are believed to have
originated as early as the Late Carboniferous or Early Permian.
Centre 4, the Mainz System, involving the Rhine-Hessen rifts,
may have been initiated much later, in the Late Mesozoic or
Tertiary period. Centre 3 may have originated in the Triassic.

Figure 9
Trough Systems around the British Isles.
(including location of seismic profiles shown on Figure 20)

The Northern North Sea Trough can be regarded as the *failed arm* of a triple junction which was centred north of the Shetlands. The other arms of this triple spreading system comprise the West Shetland Basin (the east flank of which is currently licensed and being actively explored) and the Voring Basin which lies in deep water beneath the Norwegian Sea. Both the West Shetland Basin and the Voring Basin appear to have become inactive towards the end of the Mesozoic period and a blanket of Tertiary sediments has been deposited across the troughs obscuring the deeper Mesozoic rifted structures.

The Forties triple junction (Centre 2) is not well defined and this again is partly due to a thick section of young Tertiary rocks which obscures the Mesozoic structure. The important Forties, Montrose, Piper, Ekofisk, Auk and Argyll oil fields are all associated with this system; productive horizons being found in Permian. Mesozoic and Tertiary reservoirs.

The Skagerrak triple junction (Centre 5) comprises one arm extending through the Skagerrak into the Oslo Graben; a second arm running offshore parallel to the West coast of Norway (West Norway Basin), and a third extending south-eastwards into Poland (Danish-Polish Trough). This junction may have been initiated as early as the Late Carboniferous or Permian.

Several phases of block faulting affected the North Sea trough system between Late Triassic and Tertiary times. These movements are thought to be due to the change in pattern of stresses and crustal readjustments which affected the margin of the continental plate adjacent to the newly developed and expanding North Atlantic Ocean. Marine conditions invaded the area both from the growing Atlantic Ocean and from the Tethys Sea, the latter located between what is now southern Europe and northern Africa. The North Sea suffered a general subsidence throughout most of the Tertiary period, accompanied by a gradual uplift of the surrounding landmasses of Norway, Scotland and England.

2.3   Evolution of the North Atlantic

From a study of the pattern of magnetic stripes in the North Atlantic, in conjunction with the results of the JOIDES (Joint Oceanographic Institutions Deep Earth Sampling) deep sea drilling programme, it can be established that prior to 200 million years ago (Late Jurassic times) the North Atlantic did not exist as an ocean, and that the continental plates of North America, Greenland and Western Europe were joined together to form one huge landmass (see Figure 7). Sometime during the Middle Jurassic period a constructive plate margin began to develop between the North American continent to the west of France and Spain to the east, and it was at that time an embryonic Atlantic Ocean first appeared.

During the subequent period from 200 to 100 million years ago (Mid-Jurassic to Mid-Cretaceous times) a continuous, but slow separation of the continental blocks of southern Europe and North America led to new oceanic crust being generated over the entire length of the southern North Atlantic. To the north the development of the plate boundary between the potential Greenland and North American Plates to the west and the European Plate to the east had only reached a preliminary stage during this time period, with tensional stresses in the crust giving rise to a chain of asymmetric basins and fault-bounded troughs along the west coast of Britain and Ireland.

During the Upper Cretaceous period, a rather poorly defined triple plate junction existed south of Ireland. Many of the complications of the continental margin now seen to the west of the British Isles appear to have been the result of intermittent but abortive attemps to extend the mid-oceanic spreading ridge northwards between the Greenland Plate and the European Plate. At the same time spreading was actively taking place along a north-westward arm which began to separate the North American Plate from the Greenland Plate, creating the intervening Labrador Sea.

A tongue of deep water, the Porcupine Seabight, trending northwards into the submerged continental shelf southwest of Ireland, is thought to represent one of the earliest axes of attempted northward extension; although the less well-defined trough system which follows the present-day line of the Celtic Sea troughs and Western Approaches-English Channel basins may represent an even earlier but similarly failed attempt.

Following the failure of Porcupine Seabight as a potential constructive plate boundary, the line of attempted northward extension appears to have shifted westward. Throughout much of the Upper Cretaceous it is believed to have formed an arm of active spreading along the line of Rockall Trough, partially separating the continental plate which formed Greenland and Rockall Plateau from that forming Western Europe.

Rockall Trough exhibits several primary stages in the separation of two continental plates, with most of the northern section still being floored by somewhat thinned continental crust, while southwards the continental crust gives way to oceanic crust, across which a few weak *magnetic stripes* can be detected. The entire trough system is over 1,000 miles in length, and since its initial creation as a primary structure has been partially infilled by thick Mesozoic and Tertiary sediments. Separation of Rockall Plateau from the major European continental plate along the line of the Rockall Trough appears to have ceased by Middle to Late Cretaceous times (between 90 and 60 million years ago), and the axis of active sea-floor spreading again shifted westwards by some 900 miles to finally form a successful northward extension of the mid-Atlantic ridge system between Greenland and Rockall Plateau. The spreading axis has maintained this position from the Early Tertiary to the present day. During this westward shift of the spreading axis, a minor failed spreading arm may have extended northwards across Rockall Plateau for a short time, creating the relatively poorly developed, shallow Hatton-Rockall Trough.

The pattern of well defined magnetic lineations across the floor of the North Atlantic between Rockall Plateau and southeast Greenland indicates that active sea-floor spreading has been taking place about the Reykjanes Ridge since the beginning of the Tertiary (60 million years ago). The initial break-up of the Greenland-Rockall Plate was accompanied by an outburst of volcanic activity, which culminated in the outpouring of extensive sheets of volcanic lava across much of southeast Greenland, Rockall Plateau, Northern Ireland, the Inner Hebrides and western Scotland. Volcanic activity has persisted until the present in Iceland.

Throughout the Palaeocene and Lower Eocene period, magnetic stripe data has shown that active sea-floor growth was taking place simultaneously about both the northwestern arm through the Labrador Sea, and the northeastern arm between Rockall and Greenland. However, by the Mid-Eocene period (47 million years ago) spreading had virtually ceased in the Labrador Sea, and the entire movement of the European Plate away from the Greenland-North American Plate was restricted to the North-East Atlantic System. This Lower Tertiary spreading axis forms the present day northward extension of the North Atlantic Ridge System separating the East Greenland plate margin from the Rockall Plateau-Norwegian-Spitzbergen plate margin and continues northwards across the Arctic Ocean into East Siberia.

Initially separation of Greenland from Rockall Plateau was associated with a relatively high sea-floor spreading rate (1.7 cm/year) which gradually decreased to approximately 0.7 and 0.8 cm/year throughout much of the rest of the Lower Tertiary, increasing again to rates of over 1.18 cm/year in Upper Miocene, Pliocene, Quaternary and present day times.

During the Eocene and Oligocene period, a phase of widespread subsidence followed the initial rapid Lower Tertiary separation of Greenland from North-western Europe, and Rockall Plateau along with much of the continental shelf to the north and west of the British Isles became submerged. At the same time a pattern of deep-water sedimentation was established in the previously existing depressional troughs of Hatton-Rockall, Rockall, Porcupine, the Celtic Sea and Western Approaches. Data obtained from the deep sea drilling programme in the North Atlantic by the Glomar Challenger (see Figure 19) indicates a second period of subsidence across the continental shelves during Oligocene times, while more recently, during the Early Pleistocene advance of the ice-sheet, a glacial fall in sea-level caused the shallower regions of Rockall Plateau and the shelf west of Britain to emerge from below the sea for a short period.

2.4 Tectonic Framework of Western Europe

From Precambrian times onwards the structural framework of Western Europe has been controlled by the existences of three extensive rigid blocks or *shields* of ancient basement rock, which at the present day occupy much of Russia and the Baltic region to the north, large areas of North Africa and Arabia to the south, and the remnants of a once much larger shield to the northwest in the Hebridean region of Scotland.

From time to time throughout the Palaeozoic and subsequent history of Western Europe, relative movement of these three great stable blocks has caused the intervening area of more mobile crust to be subjected to periods of strong compressive force, which in turn gave rise to the development of three major chains of mountains; firstly the *Caledonian chain* in Lower Palaeozoic times; later the *Hercynian chain* in Upper Palaeozoic times, and finally the *Alpine chain* in relatively recent Upper Tertiary times (see Figure 10).

It is now fairly widely accepted that such mountain chains represent the collision boundary between two advancing plates whereby the original intermediary marine and continental sediments become gradually more and more compressed, increasingly folded, and finally, following collision and possible underthrusting of one plate margin beneath the other, thrown up and severely contorted to form a complex system of overfolds and thrusts.

The rocks forming the ancient shield blocks are composed chiefly of crystalline Precambrian rocks overlain in part by thick sequences of relatively undisturbed younger strata, and were themselves originally involved in powerful earth-movements in earlier Precambrian times. From the beginning of Palaeozoic times onwards, these blocks have behaved as strictly rigid platforms, whose margins have acted as a constraining boundary against which younger sediments have been strongly folded and overthrust. In the north, the Precambrian floor of the Russian Platform is only exposed at the surface in the monotonously flat region of the Baltic States where it is known as the *Baltic Shield*. The western limit of this shield runs along the middle of Scandinavia (see Figure 10) forming the eastern boundary to the Caledonian mountain chain, of which only the denuded roots are preserved today.

The *Caledonian mountains* represent the Lower Palaeozoic collision boundary between the two plates comprised chiefly of the Baltic Shield to the east, and the platform occupying the Scottish Hebrides (and probably in pre-North Atlantic drift times, much of

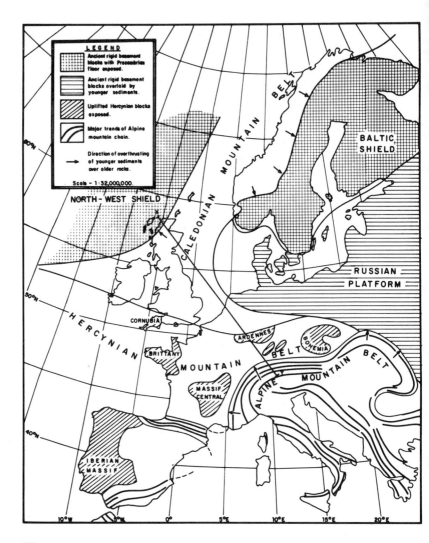

Figure 10
Structural framework of Western Europe.

The map above showing the major structural elements across Western Europe bears a strong resemblance to the present day pattern of highlands and lowlands. The Alpine structural belt, thrown up in the Tertiary, still forms a more-or-less complete mountain chain across Southern Europe, while the older Caledonian and Hercynian chains formed during the Palaeozoic have suffered strongly from subsequent erosion, so that the lower, most denuded parts, along with the southern portion of the Baltic Shield have been overlain by thick sequences of younger Palaeozoic, Mesozoic and Tertiary sediments, many of which are hydrocarbon bearing.

The diagrammatic cross-section X-Y shown in Figure 11, shows how the thickest sedimentary accumulations overlie the heavily denuded portions of the Caledonian and Hercynian mountain belts, while the ancient North-West Shield, the uplifted Ardennes block and the young Alpine mountain chain present relatively sediment-free surfaces.

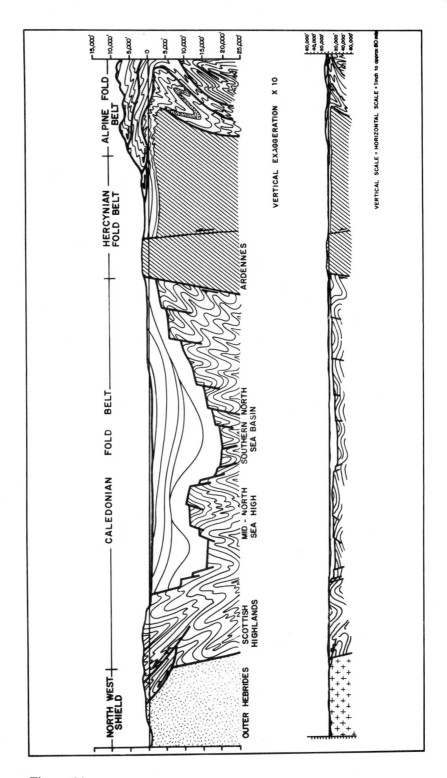

Figure 11
Structural cross-section across North-western Europe.

Greenland) to the west. At the present day the Caledonides are exposed along the length of Norway and in Scotland. However, originally the mountains formed a broad continuous belt extending from Norway, beneath the North Sea through Scotland, northern and central England, to branch southwestwards beneath Ireland, and southeastwards curving around the margin of the Russian Platform to underlie considerable parts of Belgium, Holland and possibly some of Denmark. To the south, the margin of the Caledonian chain is abruptly terminated by the cross-cutting folds of the younger *Hercynian chain*. As a result of the development of these Caledonian movements the tectonic framework of much of northern and western Britain took on a new NNE-SSW grain which has acted as a strong influence over the subsequent trend of younger tectonic structures.

About three-quarters of the entire area covered by the British Isles lies within the Caledonian fold belt, although over most of England the denuded roots of these mountains are hidden beneath younger formations and modified by younger structures. The western boundary of the fold belt with the North-West Shield is exposed as a well-defined major thrust line, the Moine Thrust, exposed in north-western Scotland, while the eastern boundary running the length of the Scandinavian peninsula is shown by strong overthrusting of the folded sediments eastwards onto the Baltic Shield.

The second phase of active mountain building to affect Western Europe, occurred as a series of tectonic pulses throughout the Upper Palaeozoic, beginning in Late Devonian times and finally dying out in the Late Carboniferous times accompanied by widespread volcanic activity. These movements led to the emergence of a roughly east-west trending chain of mountains extending from Porcupine Bank and the Celtic Sea across the Channel into central Europe. The Hercynian chain today is recognisable as a series of uplifted remnant blocks of deformed basement rocks which break through the cover of Mesozoic and Tertiary sediments across Western Europe. Piecing together the isolated remnants which form the Brittany peninsula and the Massif Central in France, the Ardennes in Belgium and the Black Forest, Vosges, Hartz, Thuringian and Eiffel mountains in central and southern Germany, it can be seen that the Hercynian fold chain was composed of two great mountain arcs (see Figure 10), the northern boundary of which cut across the Dutch, Welsh and southern Irish portion of the Caledonides, to converge westwards with the margin of the ancient North-West Stable Shield. A further group of remnants occurs to the south of the Pyrenees, over the western half of the Iberian Peninsula, as small engulfed blocks caught up in the centre of the young Alpine chain, and also to the east of the Black Forest, in the uplands of Bohemia.

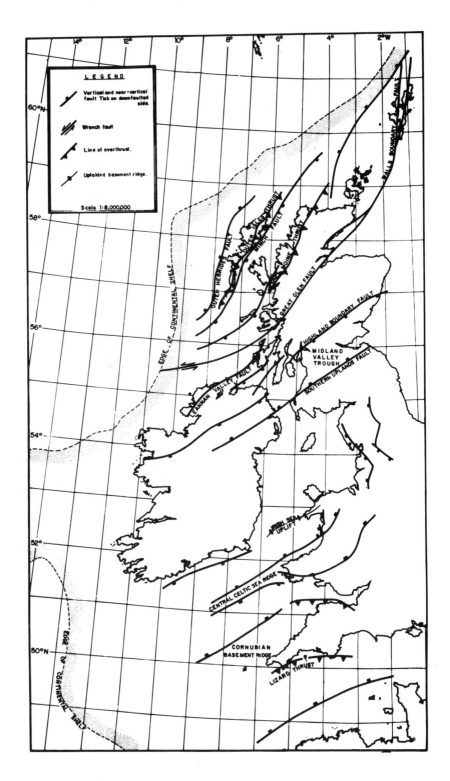

Figure 12
Major fault lineaments of West Britain

Although the main belt of intense Hercynian folding lies chiefly to the south of the British Isles, much of central and southern Britain, which already lay within the earlier Caledonian fold belt, was again affected and modified by Hercynian folding and faulting. The Hercynian tectonic movements influenced the subsequent development of younger structures in southern Britain, and to the north in Scotland gave rise to considerable lateral displacement of the crystalline Caledonian rocks along a number of northeast trending tear faults such as the Great Glen-Walls Boundary Fault and the Minch Fault.

Towards the end of the Upper Palaeozoic period, the earlier predominantly compressive regime changed to one of local extension and crustal collapse as the primary tensional stresses, pre-dating the separation of the European and Greenland-American plates and the initiation of sea-floor spreading in the North Atlantic, began to make themselves felt. A pattern of narrow, interlinked and fault-bounded continental troughs became superimposed across both the newly uplifted Hercynian fold belt and older, by then denuded Caledonian root zone, to form a series of roughly parallel features which extend from beyond the Arctic Circle in the north, down both sides of the British Isles and into the present-day Atlantic Ocean and Netherland — central Germany area respectively (see Figure 9).

In Mesozoic times a considerable part of the Hercynian fold chain became submerged beneath a thick accumulation of marine sediments which were being laid down across southern Europe in what was then the Tethys Sea. By Upper Tertiary times the third and youngest period of strong mountain building activity took place in southern Europe, contorting and throwing up these sediments into the intensely buckled and overthrust peaks of the Alpine mountains. The long sinuous line of this Alpine belt extends as a series of loops from the westernmost edge of the Mediterranean, in the vicinity of Gibraltar, through central southern Europe and eastwards across Asia Minor. This line represents the complex Miocene collision boundary between the southward moving *Eurasian plate,* composed of the rigid Russian-Baltic Platform and its attached Palaeozoic fold belts, and the northward moving and subsequently underthrust *African plate,* comprising parts of the Mediterranean floor and the rigid North African-Arabian Platform.

Although, as a result of collison, the Alpine folds were thrust northwards across the older Hercynian chain obscuring the original southern margin of this fold belt, most of the northern and central part of the Hercynian fold structures, (with the exception of the Carpathian region) still remain visible. The Carpathian section of the Alpine chain is unique in that it was thrust forward beyond the northern Hercynian fold edge to rest directly on the edge of the Russian Platform.

Outside the boundary of the mountain chain itself Alpine movements (possibly mainly of Miocene age) have determined the structure of the Mesozoic and Tertiary strata in North-western Europe, Britain and the North Sea. Folding of this age is strongly developed only in Europe, the southern part of Britain, south of the London-Brabant Massif, and there is marked decrease in intensity northwards away from these regions. However, reactivated movements along older fault lines and the development of new faults appears to be a common feature of the Tertiary everywhere.

Selected Reading

AVERY, O E et al	1969	Morphology, magnetic anomalies and evolution of the Northeast Atlantic and Labrador Sea. Trans Am Geophys Union. 50(4).
BULLARD, E C et al	1965	The fit of Continents around the Atlantic. Phil Trans R Soc (A) 258.
DEWEY, J F and BIRD, J M	1970	Mountain belts and the new global tectonics. J Geop Res Vol 75 No 14.
LAUGHTON, S L	1971	South Labrador Sea and the evolution of the North Atlantic. Nature 232 (5,313).
LE PICHON, X	1968	Sea-floor spreading and continental drift. J Geop Res Vol 73. No 12.
TARLING, D H and M P	1971	Continental drift. pubs Bell and Sons Ltd, London.

3 Channel Basin

3.1 Introduction

The Channel Basin forms a sedimentary trough beneath the eastern portion of the English Channel between the south coast of Dorset and Hampshire and the north coast of Brittany. The basin extends onshore into southern Britain and northern France and eastwards to roughly 1°E longitude; westwards it extends down the Channel to the cross-cutting Cherbourg — Plymouth basement barrier. This barrier is an uplifted ridge of Precambrian and Palaeozoic metamorphic rocks which runs north-westwards from the Cotentin Peninsula to the southern tip of Devon, separating the depositional trough of the Channel Basin from that of the Western Approaches Basin.

Structurally the Channel Basin comprises a series of WSW-ENE trending synclinal basins separated by intervening upfolded ridges which run roughly parallel to the coastline of southern England. The rocks infilling these basins range in age from Permo-Triassic to Tertiary, and represent a direct offshore continuation of the Mesozoic and Tertiary sediments found along the adjacent English and French coastline.

The stratigraphical succession exposed along the south coast of Devon, Dorset, Hampshire and the Isle of Wight is discussed in some detail as it is anticipated that a very similar sequence will be developed below the marine area of the eastern English Channel. Knowledge of the geological sequence offshore over the British sector of the Channel is drawn largely from the considerable amount of shallow seismic reflection and sea-floor sampling data that has been obtained over the last few years by University College, London, and other academic bodies. French scientists have also been active in gathering geological and geophysical data in the English Channel.

3.2 Geophysical and Bottom Sampling Data

A systematic collection of geophysical data and some 280 sea-floor rock samples over the eastern part of the English Channel was made between 1968 and 1971 and forms an easterly continatuion of an extensive work programme carried out earlier by a team from Bristol University over the Western Channel and Western Approaches.

The geophysical work yielded approximately 3,500 nautical miles of shallow seismic reflection, magnetic and echo sounding data.

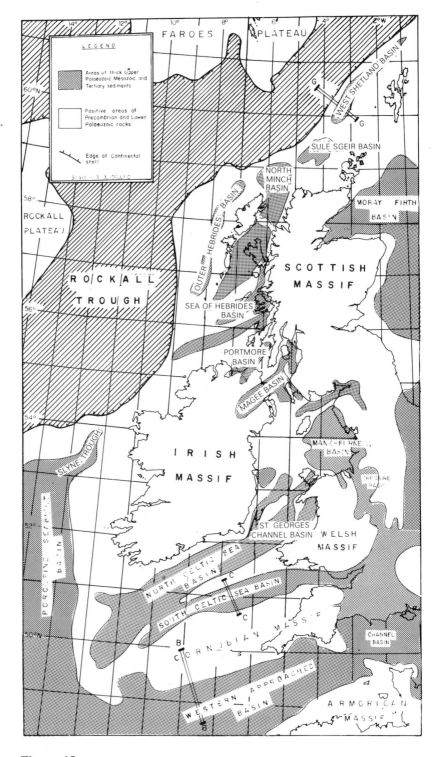

Figure 13
Sedimentary basins of West Britain
(including location of seismic profiles shown on Figure 16 and 28)

Coring of the sea bed on a 2.5 nautical mile grid was also carried out. In order to delineate the major geological structure of the basin, the grid traverses were run at right angles to the anticipated trend of the folds. Although the maximum depth to which seismic reflection information was obtained did not exceed 1,000 feet, the penetration was sufficient to be able to pick out the major structural features. In addition, the magnetic data gave further information on the variation in the depth to the underlying Precambrian and Palaeozoic basement horizon.

In the absence of any deep offshore borehole evidence or velocity measurements, correlation of the seismic reflection profiles with the geological sequence known onshore in Britain and France is based entirely on the samples of rock obtained from the sea bed. An analysis of the microfossils obtained from these core samples has also aided in the identification of the stratigraphical horizons visible on the seismic reflection records, the results of which have been used in compiling the map shown as Figure 14.

3.3   **Stratigraphy and Structure**

The structural relationship of the strongly folded deep water Carboniferous Culm rocks forming much of the basement of the Cornubian Massif, to the shallow water and less disturbed Carboniferous rocks exposed in South Wales and north Somerset is imperfectly known, as the junction is largely obscured beneath a cover of Jurassic and Triassic sediments. This cover extends westwards along the Bristol Channel into the Celtic Sea. It is believed however, that the deeper water rocks were thrust northwards across the shallow water sequence during the Hercynian phase of mountain building. The eastward extent of the folded deep water rocks below the Channel Basin is unknown, but it is conceivable that their earlier limit is marked by a zone of faults associated with the Plymouth-Cherbourg Ridge. It may be speculated that coal measures of Carboniferous age extend from their southernmost known exposures (the Somerset and Kent Coalfields) beneath the Permian and Triassic sediments of the English Channel.

The pre-Permian rocks of the Channel area are generally considered as economic basement with reference to hydrocarbon exploration. However, the distribution of the Carboniferous, and in particular, the Upper Carboniferous Coal Measures is of economic importance as the interbedded coals and shales form potential source rocks for methane as they do in the southern North Sea Basin, northern Holland in Germany.

A number of wells drilled onshore in the Wealden area of southeast England have reached the Palaeozoic basement horizon, but

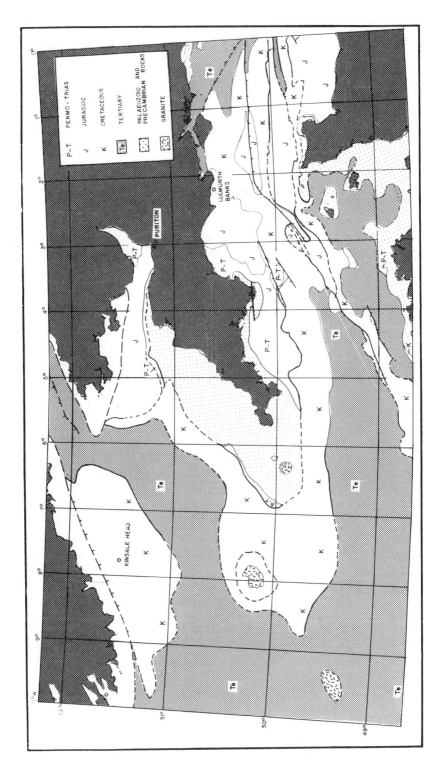

Figure 14
Geological map of the Celtic Sea, Western Approaches and English Channel.

the only proven occurrence of coal measures is in Kent, where the Carboniferous measures lie in a WNW trending syncline at depths of between 800 and 3,800 feet. Lithologically they comprise a sequence of mudstones, siltstones and sandstones with thin seat earths and coal seams, which rests unconformably on Lower Carboniferous limestones, and overlaps northwards and eastwards across Devonian and older rocks. The evidence accumulated to date indicates that the onshore Mesozoic Weald Basin is underlain chiefly by folded Carboniferous and Devonian rocks. The strikes of the folding in the Kent Coalfield is NW-SE, but further west a Hercynian (E-W) trend is considered more likely.

In Dorset and to the West in Devon, the Jurassic is underlain by a thick Permo-Triassic sequence of sands and shales in a red-bed facies. These rocks have a limited exposure on the floor of the Channel, due to the widespread cover of younger Jurassic, Cretaceous and Tertiary sediments. Where exposed onshore, the Permo-Triassic sediments appear to have been deposited in a semi-arid or desert environment and are therefore unfossiliferous. Their subdivision into Permian and Triassic horizons is based entirely on their lithologic character and stratigraphic position. Earlier authors have described this sequence as the *New Red Sandstone,* as distinct from a lithologically somewhat similar, but much older Palaeozoic sequence, the *Old Red Sandstone* (Devonian) which also developed as a red-bed facies during a period that followed active mountain building.

The upper part of the New Red Sandstone in Devon shows a marked similarity to the Triassic *Keuper Marls* and *Bunter Sandstone* developed elsewhere in Britain, while the underlying Permian bears a strong resemblance to the Lower Permian (Rotligendes) facies which is developed in West Germany.

On the Devon coast the cumulative thickness of the New Red Sandstone sequences exceed 8,000 feet and they are expected to reach thicknesses of several thousands of feet beneath the western part of the Channel Basin.

Permian breccias lie with strong angular unconformity on the folded Devonian and Carboniferous basement rocks of Devon and Cornwall. This region was uplifted during the Hercynian period of mountain building to form the large upland region of the *Cornubian Massif.* Large granitic masses were intruded into the basement at the same time, but were not exposed at the ground surface until the Cretaceous; consequently the bulk of the detritus in the breccias was derived from the limestone, sandstone and slate horizons forming the basement and the metamorphic rocks associated with the intrusions of granite.

The breccias were deposited as alluvial fans across a semi-arid

Devonian-Carboniferous land surface of considerable physiographic relief. Deep valleys controlled by the east-west Hercynian structural trend were eroded and subsequently infilled with these New Red deposits. Progressive infilling caused the ridges between the valleys to be buried, and an alluvial plain gradually extended westwards beyond the limit of the valleys. Topographic ridges some 4,000 feet in height were buried and the original relief between the floor of the New Red Sandstone Basin and the summit of the Cornubian upland was probably in excess of 12,000 feet.

The depositional basin of erosional red-bed sediments extended northward into Somerset and southeastwards under the English Channel. The eastward limit of the basin is unknown. The alluvial fans are regarded as a marginal facies to a desert basin which lay to the east. Within this desert basin, sandstones and siltstones of fluviatile origin and wind blown dune sands were deposited contemporaneously. The aeolian sands show the prevailing wind direction during the Permian to be from the south or southeast.

The breccias pass eastwards below the Channel Basin into sandstones and siltstones and are probably similar in many ways to the Rotliegendes deposits of the southern North Sea Basin. Because of the nature of deposition, the sands are likely to include porous, potential hydrocarbon reservoir rocks, the prospectiveness of which is dependent upon their relationship with either older or younger hydrocarbon source rocks.

Triassic red-beds, including conglomerates, sandstones and marls, with minor gypsum and rock salt horizons, are exposed on the Channel coast of Devon where they can be seen to dip eastwards below the Jurassic. Although the top of the sequence (Keuper Marls) has been recorded in boreholes as far east as Portsdown, in Sussex, onshore outcrop and borehole data suggests that the thickest part of the Triassic basin probably lies west of the Isle of Wight. It is supposed that a lobe of the sedimentary basin extends eastwards to underlie the Weald Basin in Sussex and Kent. The gravity low located over the Hampshire Basin may reflect the presence of an underlying thick sequence of New Red Sandstone rocks, while the marked gravity gradient across the Isle of Wight may represent a gradual southward decrease in the thickness of New Red Sandstone rocks towards the Channel Basin. Recent work suggests that the Triassic rocks underlying the Isle of Purbeck may be in the order of 4,500 feet in thickness.

The thick and porous Bunter Sandstones form attractive potential hydrocarbon reservoir beds in southern Germany, in Holland, and in the southern North Sea Basin. Similar rocks should also be considered as a potential exploration target in the English Channel. Generally the Triassic sequence lacks indigenous source rocks and is dependent upon the migration of hydrocarbons from underlying or stratigraphically younger rocks. The prospectiveness of the Lower Permian and Bunter in the Channel Basin is largely dependent upon the presence or absence of an underlying succession of Coal Measure source rocks.

A full sequence of Jurassic rocks, totalling some 4,000 feet is exposed along the south coast of Dorset (see Figure 15B) and can be seen to extend offshore into the Channel. A large area of Upper Jurassic rocks has been detected to the SSW of the Isle of Wight by shallow seismic reflection work and bottom sampling. Southwards again, Jurassic rocks are known to floor the English Channel north of the Cherbourg Peninsula, and extend onshore into France along the northern Brittany coast. A broad area of Jurassic rocks also occurs in the easternmost Channel and on the French coast south of Calais. A small area has also been discovered in mid-Channel, south of Plymouth.

It is probable that Jurassic rocks are present beneath the whole of the eastern Channel Basin, either exposed on the sea-floor or lying under a cover of Cretaceous and Tertiary sediment, and also beneath large areas of the western Channel.

The Jurassic sequence exposed in Dorset and the Weald Basin comprises a succession of more than 4,000 feet of clays, shales, limestones and sandstones and include several potential exploration targets in the Channel Basin.

Although a number of wells have been drilled in southern England to test the Jurassic prospects, only two small oilfields have been discovered to date. The Kimmeridge oilfield on the Dorset coast (50°38'N and 2°10'W) produces from a Middle Jurassic marine limestone with minor production from the overlying fractured shale. The Kimmeridge oilfield is developed in a minor fold on the steep north limb of the larger Weymouth Bay anticlinal structure. Recent drilling has, however, produced a significant discovery at Wych Farm, a few miles northeast of Kimmeridge, where the Jurassic forms a sharp anticline covered by flat-lying or gently synclinal Cretaceous and Tertiary sediments. This Lower Cretaceous zone of narrow folds extends westwards into Dorset, and although several structures have been drilled, no further production has yet been established.

Potential reservoir horizons exist in sands of Middle and Upper Liassic age, and in limestones of the Middle Jurassic, while thick marine clays and shales of Liassic and Upper Jurassic age provide excellent potential source rocks. Active oil seeps are known

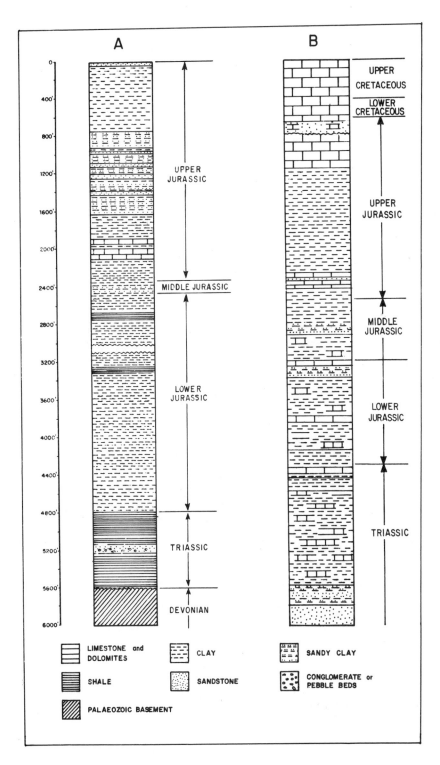

Figure 15

Generalised stratigraphy of
  a) The Bristol Channel
  b) The Dorset-Hampshire Coast.

in several localities along the Dorset coast and traces of both oil and gas have been recorded in a number of wells.

The majority of the Jurassic sands exposed onshore are fine-grained and clayey, while the limestones are similarly clayey and not very porous, providing little encouragement as potential reservoir rocks. However, the facies developments in the offshore region are as yet unknown and the development of sandy shoreline facies, similar to that known in eastern Kent, could provide more attractive reservoir horizons.

Cretaceous sediments are exposed extensively in the areas bordering the English Channel in Kent, Sussex, Hampshire, Dorset and into Devon on the British side, and from Le Havre to Calais on the French coast. Seismic reflection traverses and bottom sampling data have shown that Cretaceous rocks also floor much of the English Channel, with limited areas of Jurassic rocks occurring west of Calais, south and west of the Isle of Wight, and north of the Cherbourg Peninsula. Over a wide area between the coast of Sussex and the French coast, and also in the western Channel, the Cretaceous is overlain by Tertiary sediments.

For a number of years the stratigraphical boundary between the Jurassic and Cretaceous in southern England has been taken at the marked lithological change from clayey to sandy beds. Elsewhere in Europe, notably in the French Jura and in the Moscow Basin (where this critical part of the succession is represented by a marine sequence), the junction can be accurately defined on their marine fossil content. Recent studies of these fossil zones on the continent and their inferred equivalents in Britain have shown that throughout Europe the beginning of the Cretaceous is marked by a marine transgression and that in England the Cinder Bed at the top of the Jurassic sequence may also mark this transgression. This correlation implies that most of the highest group of the Upper Jurassic should now be considered as Cretaceous in age. The Upper Jurassic reflector which is believed to represent the high velocity limestones beneath the Cinder Bed can be mapped on the seismic records as a more-or-less continuous horizon over most of the Channel.

The Cretaceous onshore can be broadly divided into four major lithologic units: on Upper Cretaceous calcareous (chalk) unit (approximately 1,700 ft thick) resting on a thin basal succession of Upper Cretaceous clays and greenish sands, the Gault Clay and Upper Greensand (up to 225 ft thick); and a Lower Cretaceous succession of brownish-yellow sands, inaptly named the Lower Greensand (up to 575 ft thick) overlying a sequence of alternating sands and clays, the Wealden Beds, which range in thickness from a few feet to in excess of 3,000 ft.

The Wealden sands and clays were laid down as lacustrine and deltaic deposits across two slowly sinking Cretaceous depressions centred over the Weald in southeast England and over the Channel to the south of the Isle of Wight, and were subsequently overlain by marine sands and shales (the Lower Greensand) deposited during a marine transgression towards the end of Lower Cretaceous times. A second, subsequent Lower Cretaceous transgression laid down the Gault Clay and Upper Greensand succession. A third and much more widespread transgression occurred at the beginning of the Upper Cretaceous and deposited a thick blanket of chalk across the whole area. Finally at the end of the Upper Cretaceous, a phase of crustal movement (Laramide phase) caused widespread uplift and erosion of this chalk horizon prior to the deposition of the Tertiary sediments. In Hampshire, the Tertiary can be seen to rest directly on the eroded surface of the Upper Cretaceous chalk.

The Tertiary sequence exposed in the Hampshire Basin and on the Isle of Wight includes sediments of both Eocene and Oligocene age and reaches a cumulative thickness of 2,100 ft. This Lower Tertiary basin extends south-eastwards across the Channel where it forms a broad downfolded tract between the coasts of Sussex and Kent and that of northern France. No sediments younger than Upper Eocene age have been proved by the bottom sampling programme over this eastern part of the Channel Basin.

The Tertiary sediments consist of a broadly cyclic sequence of sands and shales with subordinate limestones, and suggests an environment of deposition which ranges from true marine to lagoonal, freshwater and continental. Onshore in south-eastern Britain, the Oligocene sediments have been involved in a phase of strong Mid-Tertiary folding, which undoubtedly correlates with the major Alpine movements of southern Europe. The Tertiary sediments of the western port of the Channel Basin are of a more marine nature than those further east, between southeast England and northern France but similarly show a strong Mid-Tertiary unconformity between rocks of Miocene and Eocene age suggesting that the area was also affected by Alpine earth movements.

It is not known whether the western and eastern section of the Channel Basin were physically separated during the Tertiary or whether they formed a single depositional basin which became more marine westwards. It seems probable that they became separated as depositional areas during the Mid-Tertiary Alpine movements which also separated the Hampshire Basin from the London Basin, and the London Basin from the Paris Basin.

Although both potential hydrocarbon source rocks and reservoir rocks are developed in the Tertiary, their shallow depth and generally synclinal disposition render them non-prospective within the Channel Basins.

Selected Reading

Chapter 3

DINGWALL, R G   1969   The geology of the Central English Channel.
Nature, Lond Vol 224.

DINGWALL, R G   1971   The structural and stratigraphical geology of a portion of the eastern English Channel.
Inst Geol Sci Rep 71/8.

DONOVAN, D T   1961   An acoustic survey of the sea-
and STRIDE, A H   floor south of Dorset and its geological interpretation.
Phil Trans R Soc Ser B Vol 244.

KING, W B R   1949   The geology of the eastern part of the English Channel.
Q Jnl Geol Soc Lond Vol 104.

4 Western Approaches Basin

4.1 Introduction

The Western Approaches Basin forms a deep ENE-WSW trending structural trough, which follows the line of the Western Approaches and English Channel between the Precambrian and Palaeozoic basement massifs of southwest England (the Cornubian Massif) to the north, and Brittany to the south. To the east the basin is separated from the similarly trending Channel Basin by a cross-cutting barrier of ancient basement rocks extending from Cherbourg towards the southern tip of Devon. To the west the trough terminates abruptly against the edge of the continental shelf.

Along with the Celtic Sea Troughs and Channel Basin, the Western Approaches Basin shows an unusual ENE-WSW trend in comparison with the more normal NNE trend prevalent over most of the continental shelf west of Britain and Ireland. A similar trend is observed for the troughs of the Grand Banks region, southeast of Newfoundland. Prior to the initiation of sea-floor spreading in the North Atlantic and the separation of Western Europe from North America, the continental shelf west and southwest of Ireland was continuous with that to the east and southeast of Newfoundland (see Figure 7). Therefore up to the point of separation sometime during the Cretaceous period, structural and stratigraphic conditions in the Western Approaches may have been very similar to those in the region of the Grand Banks basins off Newfoundland, although it is uncertain whether any direct sedimentary link existed between the two basin areas at that time.

During Carboniferous to Lower Permian times, a period of strong north-south compressive force (the Hercynian movements) created a broad chain of mountains through central France, Brittany, southwest England and southernmost Ireland, uplifting and tightly folding the Carboniferous and older rocks within this WNW-ESE zone. It seems likely that this relatively stronger belt influenced the direction of tensional stresses during the subsequent phase of Mesozoic rifting and final rupture of the North American continent away from Western Europe, leading to the creation of a triple plate junction between the Newfoundland shelf and the Western Approaches-Celtic Sea shelf in the Cretaceous and the development of a new north-western arm of spreading through the Labrador Sea. The Western Approaches trough is, therefore, thought to have formed as a Late Palaeozoic-Early Mesozoic (Permo-Triassic) rift structure subsequently infilled with thick Tertiary, Cretaceous, Jurassic and Permo-

Triassic sediments. The southern margin is strongly faulted against the Precambrian and Palaeozoic basement of the Brittany Massif, while the northern margin also shows well developed local faulting.

During the Permo-Trias, the fault-controlled basins southwest of the British Isles became centres of deposition of thick red-bed and evaporite sequences derived from the progressive erosion and destruction of the Hercynian mountain chain. Deep valleys draining from the surrounding uplands converged into the Western Approaches and Channel Basins, while across the floors of the basins themselves semi-arid and desert conditions prevailed with sands and pebble beds being deposited both as dunes, and in outwash delta fans and flood channels. This pattern of continental deposition continued until Upper Triassic times, by which time the original Hercynian highlands had become so worn down that only finer sediments such as river-deposited red mudstones and siltstones, and lacustrine salts and evaporites were being laid down across the lowland plains. Salt deposits are known to exist widely in the Triassic beds throughout Britain, and within the vicinity of the Western Approaches, have been found in Somerset, in the Cheshire Basin, and as indications on geophysical records across the Celtic Sea and southern Irish Sea.

At the end of Triassic times (Rhaetic) the sea is thought to have invaded the downfaulted depression from both the north and southwest, thus establishing a marine link with the opening Atlantic Ocean. It is probable that the entire Western Approaches area, including the adjacent land areas of France and southern England became submerged beneath a partially enclosed shallow sea in which Jurassic, the *Brabant Massif* formed a positive island feature which extends from London into northern France, although it may have been submerged during Liassic and Kimmeridgian times.

In Jurassic times deposition in the Western Approaches Basin was influenced by a number of adjacent structural elements: the Cornubian Massif, the Brittany Massif, and the Western Approaches Massif. Towards the end of the Jurassic tectonic movements (Late Kimmerian fold phase) led to uplift, faulting, folding and erosion and the general emergence of the areas, causing the Late Jurassic sea to withdraw northwards to Lincolnshire and Yorkshire and southwards to Spain. The Lower Cretaceous was marked by a return to continental and marginal marine conditions, and thick sequences of deltaic, lacustrine and coastal swamp deposits were laid down in the isolated depressions of the Channel, Western Approaches and Celtic Sea Basins. Limited sea bottom coring of the Western Approaches has shown the Lower Cretaceous to consist dominantly of continental beds.

A late Lower Cretaceous phase of crustal disturbances gave rise to a period of strong folding and faulting, the effects of which are well marked in the English Channel. Renewed marine transgression at the end of the Lower Cretaceous gave rise to local overstepping of the older continental beds by younger semi-marine and marine sediments.

The Upper Cretaceous started with a widespread marine transgression which resulted in the submergence of many of the platforms previously established by the Hercynian period of mountain building. This epoch was marked by a rapid deepening of the whole region and the deposition of a fine grained deep marine limestone known as the *Chalk*. The transgression submerged the upstanding Irish, Western Approaches and London-Brabant Massifs, and the Cornish and Welsh Massifs were reduced to islands rimmed by shoreline sediments which passed rapidly seawards into limestones. A similar relationship occurred along the northern margin of the Brittany Massif. Chalk deposition continued from the Upper Cretaceous into earliest Tertiary times.

A pre-Eocene period of earth movements, the *Laramide phase*, gave rise to renewed regional uplift and re-emergence of the platform areas. This was followed by widespread erosion which resulted in the partial removal of the Upper Cretaceous Chalk prior to the deposition of the Eocene sediments. During the Eocene the continental margin was again submerged and the sea periodically invaded the present Western Approaches Basin. As a result shallow marine sands and clays interfinger with fluviatile and deltaic sediments and freshwater limestones. The Plymouth-Cherbourg Ridge may have also been rejuvinated and uplifted during the Laramide earth movements, and thus acted as a barrier between the Western Approaches and Channel Basins from Eocene times onwards.

Uplift again occurred in the Oligocene resulting in a major retreat of the sea, and a period of erosion. In the Western Approaches and Celtic Sea, the coastal zones bordering the uplifted platform areas were the sites of coastal swamps and lake deposits, while within the Western Approaches Basin itself, freshwater limestones were deposited.

Downwarp of the continental margins occurred in the Miocene with subsidence and marine invasion along the axis of the Western Approaches Basin. The Miocene sea deposited fine grained sandy and calcareous rocks, containing an abundance of microscopic shells, across the basin. Folding and faulting related to the Alpine earth movements was accompanied by severe erosion and the removal of Miocene sediments in both the Western Approaches and the Channel. This was followed by a widespread marine transgression in the Pliocene with the sea invading from the southwest and south, as in Miocene times. Since this period subsidence has continued along the Western Approaches Basin axis into Recent times.

The land masses adjacent to the western portion of the English Channel and the Western Approaches are composed entirely of Lower Palaeozoic and older rocks (considered as economic basement for hydrocarbon exploration) and therefore, these in themselves provide no indication of the younger sediments likely to infill the intervening offshore basin. A considerable amount of shallow seismic data and sea-floor sampling work has been carried out by a number of academic institutions over the Channel, but the geophysical work does not extend in detail into the westernmost part of the Western Approaches. This poor geological control coupled with the lack of any deep boreholes within the offshore region has resulted in the geology and structure of the Western Approaches Basin being only rather sketchily known.

4.2 Geophysical and Bottom Sampling Data

Early geological work in the Western Approaches and western English Channel area was carried out by King and summarised by him in a 1954 publication. In 1957 a team from Bristol University began a systematic study of the area, and this has continued up to the present day. The Bristol group used a basic bottom sampling grid of 10 minutes longitude and 10 minutes latitude (10 x 7 nautical miles), but in areas of geological complexity a much closer spacing was employed. About 60% of the 1,600 stations have yielded useful information, and some 2,400 track miles of continuous seismic profiling has been collected.

Published seismic reflection profiles by the universities has recognised two main structural units within the Western Approaches Basin. The upper unit is formed of essentially flat-lying or gently folded sediments which rest uncomfortably on the eroded surface of a strongly folded and faulted lower unit. An analysis of seismic velocities over the basin suggest that the upper unit is composed of a Permo-Triassic-Jurassic-Cretaceous-Tertiary sequence, while the lower unit shows markedly higher velocity values and it is believed to represent folded and metamorphosed Palaeozoic basement rocks. Deep hollows in the basement surface occasionally contain a thin high velocity layer, which could be either Basal Permian or Uppermost Carboniferous in age.

The cross-section shown in Figure 16 A is based on a seismic profile across the basin, but the age identification of the different seismic layers picked out are only very tentative. In the last few years a number of oil companies have carried out extensive geophysical work across the Western Approaches Basin in preparation for the next round of licencing.

One of the problems in the identification of the subsurface seismic markers within this basin, is the lack of any nearby stratigraphic control. The nearest exposures of Permian, Mesozoic and Tertiary sediments occur along the English Channel and Bristol Channel coasts, (Figure 15) but these unfortunately represent the marginal sequences of the Channel and Southern Celtic Sea Basins respectively, and as such may show little resemblance to sediments of the same age deposited in the Western Approaches Basin. Of greater relevance to any interpretation of the sedimentary infill is the coastal and offshore stratigraphy of the Iberian Peninsula and Eastern Canada, since prior to continental break-up in the Late Jurassic and Early Cretaceous, these regions lay adjacent to the Western Approaches. For this reason, oil geologists have watched the results of drilling in the Grand Banks region of the East Canadian continental shelf with great interest.

4.3 Stratigraphy and Structure

The nature of the basement which underlies the complexly downfolded trough of the Western Approaches Basin is believed to be similar to the Precambrian and Palaeozoic rocks which flank the trough and form the Cornubian Peninsula to the North and the Brittany Massif to the south.

The Cornubian Massif extends south-westwards to the continental shelf edge beneath an increasingly thick sedimentary cover. The granites of Devon and Cornwall form part of a string of intrusive granitic bodies which include those of the Scilly Isles and the buried granites further offshore which have been detected by sample and gravity measurement methods. The Palaeozoic rocks of Cornwall are folded and metamorphosed and it is doubtful whether they could have functioned as source rocks for hydrocarbon generation after being affected by Hercynian earth movements in the Carboniferous. If these rocks are characteristic of the basement extending beneath the whole basin, then it is unlikely that the Carboniferous Coal Measures would have ever acted as source rocks for the overlying Permian and Triassic sandstone reservoirs.

The Western Approaches Basin is fringed by exposures of Permian and Triassic sediments. To the north, around the margins of the Cornubian Massif these form a red-bed sequence which passes conformably upwards into the Jurassic sediments.

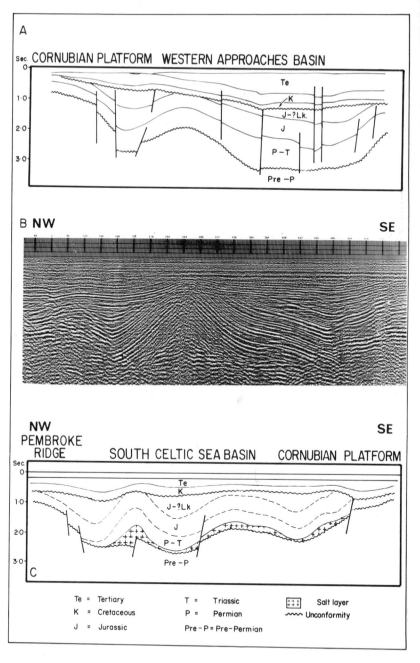

Figure 16

A   Section across the Western Approaches Basin based on an existing seismic profile.

B   Seismic profile across the Southern Celtic Sea Basin.

C   Interpretive section based on the profile above.

For locations see Figure 13

To the south a similar red-bed sequence containing a number of salt horizons occurs in Portugal. To the east, salt of Late Triassic to Middle Jurassic age has also been recorded in a deep well within the Grand Banks region. It therefore seems likely that a thick Permo-Triassic red-bed sequence, possibly containing salt horizons, will occur at depth in the Western Approaches Basin.

Sea-bed sampling has shown that rocks of Jurassic age have only a limited exposure in the Western Approaches, although they may have a widespread distribution beneath the cover of Upper Cretaceous and Tertiary rocks. Geophysical evidence suggests that between 4,000 and 6,000 feet of Jurassic rocks may be present over much of the offshore basin, although these thicknesses may be exceeded in local downfaulted trough regions of the basin floor.

Figure 15 A shows the sequence established within the Bristol Channel, and when compared with the Jurassic sequence of the south coast of England, shown in Figure 15 B, it can be seen to differ markedly in its conspicuous lack of any carbonate section. Although the Jurassic sequence in the originally adjacent areas of Portugal and the Grand Banks (Figure 7) are characterised by thick carbonate sections, it is uncertain whether similar thick carbonates typify the Jurassic in the Western Approaches Basin. The available seismic reflection data suggests that within the centre of the basin a thick argillaceous sequence comparable to that found to the north in the Bristol Channel Basin, and in the Mochras Borehole on the edge of the Cardigan Bay Basin may be more likely. However, if carbonates are present, (even towards the margins of the basin), they could act as a good potential reservoir horizons for the accumulation of hydrocarbons. Onshore in southern England, good source rocks, in the form of black, organic-rich shales are present within the Lower Jurassic section, and it is anticipated that these or similar horizons may well extend offshore into the main basin.

Although Lower Cretaceous rocks only occur in limited exposures on the sea-floor of the Western Approaches, they are widespread elsewhere around the Atlantic, as in Portugal, Eastern Canada, and further east in the Channel Basin. It seems likely, therefore, that they will similarly be present, at least locally, as a thick sequence of non-marine sandstones and shales within the Western Approaches Basin.

The Upper Cretaceous chalk sea spread over a much wider area by comparison with the restricted extent of the Lower Cretaceous basins. Nevertheless it is probable that part of the Cornubian and Brittany Massifs continued as islands even at this time. The chalk deposits of Upper Cretaceous age thicken westward into the main basin and towards the continental shelf edge although

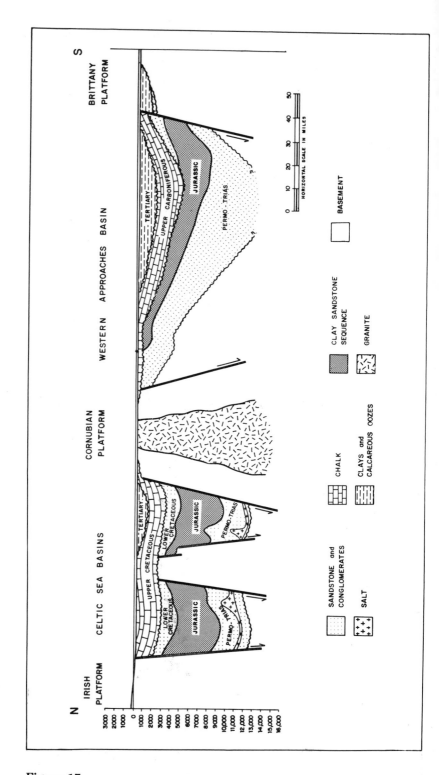

Figure 17

Diagrammatic section across the Celtic Sea and Western Approaches Basins.

the most complete sequence probably overlies the central axis of the trough.

Tertiary sediments cover much of the western part of the Western Approaches, but in general the sediments have suffered insufficient burial to be of any interest for hydrocarbon exploration.

Selected Reading

CURRAY, D et al   1970   Geological and shallow subsurface geophysical investigations in the Western Approaches to the English Channel.
Inst Geol Sci Rep 70/3.

DAY, A A et al   1956   Seismic prospecting in the Western Approaches of the English Channel.
Q Jnl Geol Soc Lond Vol 112.

KING, W B R   1954   The geological history of the English Channel.
Q Jnl Geol Soc Lond Vol 110.

5 Celtic Sea Basins

5.1   Introduction

Between the coast of southern Ireland, South Wales and the southwest peninsula of Devon and Cornwall, the marine region of the Celtic Sea is underlain by a pair of NE-SW trending, roughly parallel, downfaulted troughs which straddle both the United Kingdom and Irish continental shelves. These troughs of young sediments are separated from the deep ocean floor of the North Atlantic to the west by a now submerged plateau of basement rocks, and from each other by a central ridge which forms a south-westwards extension of the ancient basement rocks exposed in the Pembrokeshire peninsula (see Figure 14). Geological and geophysical data suggest that the troughs form partially enclosed depressions let down by faulting into the basement surface and are infilled with sediments which range from Permian or Triassic up to Lower Cretaceous age. Younger Upper Cretaceous and Tertiary sediments rest unconformably across older rocks, transgressing both the infilled basins and the eroded surface of the surrounding basement platforms.

The broader Northern Trough extends from near the continental shelf edge (10°30′W) north-eastwards close to the Cork-Wexford coastline, through St. George's Channel and links with the deep Mesozoic-Tertiary basins of Cardigan Bay and the southern Irish Sea. The narrower Southern Trough is confined between the shallow basement platform of the Cornubian Massif, and the Pembrokeshire basement ridge and similarly stretches north-eastwards into the British sector of the Celtic Sea before swinging East-West to follow the line of the Bristol Channel and the exposed downfolded sequence of the Glastonbury Syncline in Somerset (see Figure 14). The Cornubian Massif forms a broad uplifted platform of basement rocks extending south-westwards through Devon and Cornwall and offshore through the Scilly Isles towards the continental shelf edge. The Cornubian Massif, in common with the surrounding land masses of Ireland and Wales and the submerged Western Platform, is dominantly composed of tightly folded metamorphosed Palaeozoic rocks. Having lost much of the thin cover of younger rocks they may have originally had, these old massifs offer little encouragement for hydrocarbon exploration. However, remnants of Mesozoic sediments exposed onshore in the vicinity of Cardiff, South Wales, and in parts of Devon and Somerset, coupled with geological and geophysical work across the Celtic Sea and the adjoining Southern Irish Sea, Bristol Channel and Western Approaches, indicate that both the Northern and Southern Celtic Sea Troughs were sites of prolonged deposition from Late

Palaeozoic (Carboniferous — Permian) times until the present day.

Along with the Western Approaches sedimentary basin, the Celtic Sea Troughs show an unusual ENE-WSW trend in comparison with the more normal NNE trend prevalent over most of the shelf west of Britain and Ireland. A similar trend is observed for the troughs of the Grand Banks region, southeast of Newfoundland. Prior to the separation of North America from Western Europe and the opening up of the North Atlantic in the Cretaceous, the Celtic Sea Troughs, like those of the Western Approaches and English Channel (to the south of the Cornubian Massif), are believed to have been structurally related to the deep troughs of the Grand Banks region, southeast of Newfoundland (see Figure 7). Although it is unlikely that any direct sedimentary link existed between the two areas, it is thought that the pre-Cretaceous stratigraphy of the Celtic Sea Troughs may closely resemble that encountered in the Grand Banks basins.

Following the initial uplift of the Hercynian mountain chain across northern France and South-West England in the Late Palaeozoic, the adjacent foreland area to the north was subject to further vertical movements leading to the partial collapse of certain fault-bounded zones, and the establishment of a Celtic Sea basin along with many of the other Permian basins which form a system of rifted troughs across the continental shelf west of Britain. During this period the new basin in the Celtic Sea formed a single complex trough confined to the north, east and south by the high Palaeozoic basement plateau of southern Britain and Ireland, and to the west by Newfoundland and Eastern Canada. Even in its earliest states, the basin was partly subdivided into a northern and southern trough by an uplifted central basement ridge, and throughout the Permian and Triassic period, the entire trough became a site for the thick accumulation of scree, river and lake deposits derived from the rapid erosion of the surrounding high mountains and more latterly, for shallow marine lagoonal deposits.

During the Late Carboniferous and Permian period accompanying the Hercynian phase of mountain building, a line of granites were intruded into the basement rocks of Devon, Cornwall, the Scilly Isles and the offshore portion of the submerged Cornubian Massif. Offshore a large granite intrusion has been identified to the west of Lands End at 50°10'N, 7°55'W (the Haig Fras Granite), and still further southwest at 48°54'N, 8°40'W, and 49°25'N, 9°25'W. Granite masses are thought to be present buried beneath 6,000 ft and 2,000 ft of younger sediments respectively.

At the beginning of the Jurassic period, although separation of Spain from North America was not sufficiently advanced for any direct oceanic influence to be felt to the north, it appears that most of the low relief regions of this northern continent were invaded by a pre-drift shallow shelf-sea which extended between Britain and North America towards Greenland and the Arctic. Whether this sea was able to cross the uplifted basement platform separating the Grand Banks basins to the southwest from the almost landlocked Celtic Sea basin, however, is open to question. During the same period, a second shallow marine invasion originating in the Mediterranean area (Tethys Sea) to the southeast was affecting southern England and the North Sea, and this may have transgressed westwards through the narrow Bristol Channel passage to affect the subsiding Celtic Sea trough.

Towards the end of the Jurassic, a period of uplift (the Late Kimmerian phase) elevated much of Britain and Western Europe, consolidating the eroded remnants of the Armorican mountain chain into a continuous barrier which once more stretched from the Cornubian Massif to Poland. In the Celtic Sea trough this caused the Jurassic sea to slowly retreat eastwards through the Bristol Channel passage leaving shallow-marine and eventually shoreline and lagoonal conditions in the basin. The Lower Cretaceous period was marked by a gradual return to a continental environment of deposition throughout Britain and the deposition of coarse continental deposits within the Celtic Sea basins.

By the beginning of the Upper Cretaceous, the mid-Atlantic spreading ridge had extended northwards to approximately the same latitude as southern Ireland and as the effect of this new oceanic influence began to be felt, the Western Approaches, Celtic Sea, Southern Irish Sea region, in common with much of southern and eastern Britain, became submerged beneath a rapidly advancing shelf-sea. Only parts of the higher ground in Ireland, southwest and central Wales, and the Cornish area of the Cornubian Massif remained as islands above sea-level. This shelf sea deposited thick sequences of chalk across the eroded surface of the underlying Jurassic and Lower Cretaceous rocks in the deepening Northern and Southern Celtic Sea Troughs, and as a thinner, sometimes discontinuous, layer across the now submerged basement of the dividing ridge, the basement platform to the west, the southern coast of Ireland to the north, parts of the Cornubian Massif to the south, and the coastal strip of South Wales, Gloucestershire and Somerset to the east. Much of this thin chalk layer overlying the basement platforms has been subsequently removed by erosion during Tertiary and Recent times, and now all that remain onshore are a few scattered remnants to indicate the extent of the once very extensive chalk seas.

The margins of the thick chalk sequence infilling the Celtic Sea troughs appears to terminate abruptly against the edge of the

Irish basement platform, suggesting that marginal faulting was actively accompanying the rapid deepening of the central axis of the trough during the Upper Cretaceous. From Upper Cretaceous times through into the Lower Tertiary (Palaeocene — Lower Eocene) period the same chalk sea environment of deposition continued without interruption and the chalk deposits pass upwards into a younger Tertiary sequence of deeper marine sediments.

Within the Celtic Sea, Tertiary sedimentation is now only represented in the Northern Trough by a small remnant basin lying to the South of Waterford and by deposits across the central and western region of the Southern Trough. However, like the underlying chalk, the Tertiary beds may have originally extended as a far more widespread deposit.

5.2 Geophysics and Sea-Floor Sampling

Although the onshore outcrops on the landmasses surrounding the Celtic Sea give little indication of the substantial section of young rocks lying offshore, published geophysical data between southern Ireland and the Cornubian peninsula and the results of extensive commercial seismic reflection work, in conjunction with offshore drilling, show that within the fault-bounded basins of the Celtic Sea there is a thick sequence of Late Palaeozoic to Tertiary rocks.

Published seismic data across the area of the Celtic Sea itself is limited, but the results of surveys in the adjacent marine areas of the Southern Irish Sea, Bristol Channel and Western Approaches are of great relevance to any interpretation of the sedimentary succession likely to be encountered. As it is now fairly widely accepted that the continental shelf off the east coast of Canada originally lay adjacent to the Celtic Sea shelf in pre-Cretaceous times, the available seismic and drillhole data across the Grand Banks Basins have also been considered in the interpretation of the pre-Cretaceous rock succession.

An analysis of geophysical data on both sides of the Atlantic suggests that the infilling sedimentary section in the Celtic Sea Troughs is well in excess of 10,000 ft in thickness and many even exceed 20,000 ft in places. Further correlation suggests that these young Mesozoic and Tertiary sediments forming the infill are composed of three medium to low velocity layers: an upper Tertiary-Upper Cretaceous layer, an intermediate Lower Cretaceous to Permo-Triassic layer, and a lower, Carboniferous to Permo-Triassic layer, which rests on high velocity crystalline basement. A similar breakdown of the sedimentary infill is also evident from published geophysical data in the St George's

Channel Basin which forms the northeastward extension of the Northern Trough. To the east, geophysical and bottom sampling data show a downfolded basin of thin Triassic and thick Jurassic rocks preserved in the Bristol Channel between the Upper Palaeozoic basement of the Welsh and Cornubian Massif.

Sparse bottom-sampling across the Celtic Sea has established the widespread occurrence of Upper Cretaceous chalk deposits on the sea-bed.

5.3   Stratigraphy

The high velocity seismic basement which underlies the Celtic Sea troughs and forms the enclosing submerged platforms is believed to be an offshore extension of the pre-Permian rocks exposed on the surrounding land masses. In southern Ireland this basement comprises a thick sequence of metamorphosed Devonian and Carboniferous sediments caught up along the edge of the Hercynian mountain chain. The sediments include not only limestones, shales, siltstones and sandstones but also a sequence of locally productive coal measures in the Upper Carboniferous. During the Hercynian period of folding and uplift, both the carbonate and non-carbonate sequences lost their primary porosity as they became converted into hard marbles, shales and pressure-welded sandstones. As a result, this destroyed their potential as reservoir rocks, although the coal measures may still possess good source rock properties for the generation of hydrocarbons. Southwards, towards the core of the Hercynian mountain chain, the basement rocks of the Cornubian Massif become progressively more and more strongly folded and metamorphosed, and have been intruded by a number of granite bodies.

Permo-Triassic sediments similar to those believed to form the basal section in the Celtic Sea troughs fringe the margins of the Celtic Sea, and are exposed as local outcrops in South Wales, Gloucestershire, Somerset and Devon. In Devon, around the northern, eastern and southern margins of the Cornubian basement massif, up to 8,000 ft of continentally deposited red siltstones and sandstones rest directly on the basement surface, and pass upwards into the Jurassic sequence.

East of the Celtic Sea, Permo-Triassic rocks occur in the Bristol Channel and onshore infilling a narrow but deep basin between the Quantock Hills and Mendip Hills. This basin forms an eastern extension of the Southern Celtic Sea Trough. The Bristol Channel at the time of Permo-Triassic deposition, was only a shallow trough linking deeper basins to the east and west, and geophysical and sea-floor sample data, shows it to be infilled by as little as 800 ft of Triassic sandstones and calcareous siltstones. However, to the east in Somerset, the Puritan Borehole (Figure 14) encountered over 1,200 ft of interbedded Upper Triassic salts and marls resting on a further 1,000 ft of Lower Triassic and Permian coarse continental sandstones and conglomerates (pebble beds). Further evidence for thick salt deposits laid down in the Permo-Trias is suggested by seismic profiles across the St George's Channel Basin west of Pembrokeshire (see Figure 22).

Along the north coast of the Bristol Channel in the vicinity of Cardiff and west Gloucestershire, Permo-Triassic sandstones and pebble beds rest unconformably on the folded surface of the Devonian and Carboniferous basement. East of the Welsh Massif, thick Permo-Triassic continental sediments fill the Worcester and Cheshire Basins (described in Section 8) and again to the north, borehold data and sea-floor sampling also show the Irish Sea to be underlaid by a thick Permo-Trias section. On this evidence it seems certain that a similar Permo-Triassic continental and shallow marine section, possibly containing thick salt horizons, exists at depth in the Celtic Sea Troughs. Although the total thickness of this section is difficult to estimate, it seems likely that it may be in the order of several thousand feet.

Direct evidence of Jurassic rocks from sea-bed coring is lacking over most of the Celtic Sea due to the extensive surface cover of Cretaceous and Tertiary sediments. Jurassic rocks are, however, well documented in southern England where they form a continuous series of exposures from Dorset through Somerset and the Bristol Channel into southeast Wales. In general, exposures in southern England (Figure 15A) show the Lower and Middle Jurassic to be dominated by marine clays and limestones interrupted by the occasional shallow marine sandstone horizon, and the Upper Jurassic by a thick sequence of deep marine clays broken occasionally by thin limestone bands.

In the Bristol Channel the Jurassic rocks infill a westward plunging synclinal basin. The basin follows the East-West trend of the Channel and is strongly asymmetrical with a gently dipping northern fold limb partially exposed in South Wales, and a steeply-dipping, partly faulted, southern limb which abruptly terminates against the underlying Devonian basement rocks off the Devon-Somerset coast. Sea-floor coring over the Bristol Channel has established the presence of Lower, Middle and Upper Jurassic calcareous and sandy clay sequences similar to those known from nearby onshore exposures (Figure 15B). These Jurassic rocks pass westwards beneath a thickening cover of Cretaceous and Tertiary sediments which suggests that they may well be widespread and thickly developed beneath the Cretaceous unconformity in the Southern Celtic Sea Trough.

Jurassic rocks have also been proved on the eastern margin of Cardigan Bay in the Mochras Borehole, where a monotonous section of Lower Jurassic mudstones and siltstones measuring 4,282ft has been encountered. Geophysical data over the southern Irish Sea suggests a similar section also lies at depth in the St George's Channel Basin and in all likelihood in its southwestern extension, the northern Celtic Sea Basin too. Although the existence of a Middle and Upper Jurassic section is still only conjectural here, it seems likely in view of the considerable thickness of Middle Jurassic rocks proved elsewhere in the southern Irish Sea.

At the present time all the evidence from the areas adjoining the Celtic Sea point to the probable development of a complete, gently folded, Jurassic sequence at least 2,000ft thick, and possible considerably more, infilling the deep Celtic Sea Troughs beneath the much more flat-lying Cretaceous strata. Jurassic sediments may have originally extended beyond the limits of these troughs to overlie the marginal basement platforms and the central dividing barrier, but if so, the sequence here will not only be markedly thinner, but will almost certainly be incomplete. It can be seen however, that much of what can be said regarding the distribution and nature of the Jurassic rocks in the Celtic Sea basins is conjectural or deductive — a feature of much of the geology of offshore West Britain. Only few oil companies have deep borehole information in the area.

Cretaceous rocks form much of the sea-floor of the Celtic Sea and stratigraphically can be divided into two units of very different lithology; the Lower Cretaceous consisting predominantly of sandstones and shales, and the Upper Cretaceous of chalk. Although Lower Cretaceous rocks are poorly exposed around the margins of the Celtic Sea it is generally believed that these areas, along with the Cornubian Massif to the south and the basement platform to the west, were strongly uplifted during the Late Kimmerian period of movement. As such, these regions were subject to rapid erosion which in turn led to the deposition of thick sequences of deltaic sands and shales in the intervening Celtic Sea and St George's Channel structural depresssions. Similar sites of localised deposition are evident to the east over southern England, where up to 3,000 ft of deltaic sands and shales were laid down, and to the west, on the other side of the Atlantic where a similar sequence, incorporating thick bodies of sand up to 1,500ft in thickness, have been encountered in the then adjacent basins of the east Canadian shelf.

It is highly probable that the gas discoveries in the Northern Celtic Sea Basin off Kinsale Head originate from similar Lower Cretaceous deltaic sandstone reservoirs. Although the precise nature of the Lower Cretaceous is still speculative here, this sequence, along with the Jurassic and Triassic sandstone horizons, represents one of the prime targets for hydrocarbon exploration in both the Northern and Southern Celtic Sea Troughs.

Unlike the confined area occupied by Lower Cretaceous deposits, the Upper Cretaceous chalk was widely deposited over Northern Europe, the English Channel, southern England and the Celtic Sea. Characteristically it is a white, fine-grained limestone almost free from land derived materials and periodically interrupted by layers of well developed black chert nodules. The Chalk extends over the basin and surrounding basement platforms alike, resting more or less conformably on the Lower Cretaceous deltaic series infilling the structural depressions and unconformably on the eroded basement surface of the denuded margins. Although at the present day, the northern limit of the Chalk lies just off the south Irish coast, close to the margin of the deep North Celtic Trough, it is evident from an Upper Cretaceous exposure preserved onshore within a collapsed cave system at Ballydeenlea, near Killarney, that chalk was originally deposited as a thin layer over much of the South Ireland platform.

Chalk is thickly deposited over the uplifted barrier separating the two Celtic Sea Troughs and across the Cornubian Massif to the south, although the highly glauconitic and sandy nature of the basal chalk exposed in Devon and Cornwall, indicates the proximity of a shoreline environment and suggests that at least part of the Cornubian Massif was emergent as an island feature in early Upper Cretaceous times. In southern England, the Chalk attains a maximum thickness of 1,750ft and similar thicknesses have been indicated by seismic profiles offshore in the Southern Irish Sea, where up to 2,000ft of Chalk rest on a probable shoreline sand-clay sequence some 700ft thick, and in the Western Approaches, where 1,650 ft of Chalk rest on the Lower Cretaceous sands and shales. The thickness of the Chalk underlying much of the Celtic Sea is not known, but it is likely that within the stuctural basin area thicknesses at least equivalent to those found elsewhere in southern England will exist.

Sea-floor sampling and seismic data over the Celtic Sea show that Tertiary sediments are not widely distributed, and although present as a westward thickening sequence of marine clays near the continental shelf edge, they are largely absent to the East over the submerged Cornubian Massif and large parts of both Celtic Sea basins. However, Tertiary sediments do occur (Figure 14) as a tongue-like feature, stretching through St George's Channel across the central uplifted basement ridge to overlie parts of the Southern Trough and also as an isolated ENE-WSW trending narrow basin over the Northern Basin. The sediments are chiefly composed of Palaeocene and Eocene deposits only a few hundred feet thick. In these areas chalk deposition appears to have continued into the Palaeocene without a break and similarly the Eocene includes a high proportion of unconsolidated marine carbonate rocks and soft calcareous clays.

Selected Reading

AUSTIN, G H   1973   Canada offshore.
                                   Amer Asso Petr Geol Bull
                                   Vol 57 pt 7.

AYRTON, W G   1973   Regional geology of the
BIRNIE, D E   Grand Banks.
                                   Bull Canad Petroleum Geol
                                   Vol 21 No 4.

BURNE, R V   1973   Palaeogeography of South
                                   West England and Her-
                                   cynian continental collison.
                                   Nature Phy Sci Vol 241 Feb
                                   12.

DONOVAN, D T et al   1973   The geology of the Bristol
                                   Channel floor.
                                   Phil Trans Roy Soc Lond (A)
                                   Vol 274.

6 Rockall Plateau and Trough

6.1 **Introduction**

Rockall Plateau forms a well-defined shoal area some 300 miles west of the Scottish and Irish coast (see Figure 18). It has an area of about 93,000 sq miles, as defined by the 1,000 fathoms isobath, and is separated from the submerged continental shelf west of the British Isles by the 1,500 fathom deep Rockall Trough.

There is now general agreement that prior to the attempted northward extension of the mid-Atlantic spreading ridge through Rockall Trough in Late Jurassic to Early Cretaceous times, Rockall Plateau was attached to both the European continent and the North American-Greenland continent. Following rupture, the Plateau became laterally displaced from West Britain by rifting and limited sea-floor spreading along the axis of Rockall Trough, and subsequently in Lower Tertiary times became separated from Greenland by active spreading about the Reykjanes Ridge. As a result the pre-Cretaceous geology of the Plateau closely resembles that of both southeast Greenland and the continental shelf west of Scotland and Ireland, and it is thought to be composed almost entirely of metamorphosed Precambrian basement rocks overlaid and intruded to the north by Upper Cretaceous to Lower Tertiary igneous rocks, none of which are suitable for the generation or preservation of hydrocarbons.

Geophysical data show that the intervening deep-water Rockall Trough is partially infilled by some 10,000 ft of sediments which rest on a basement horizon of consolidated Lower Mesozoic or Upper Palaeozoic sediments in the central and northern section of the Trough, and possible Upper Mesozoic (Late Jurassic to Mid-Cretaceous) oceanic crust in the south.

The lack of drillhole data across Rockall Trough makes any identification of the age of the infilling sedimentary horizons still largely guess-work. To date, the oldest sediments to be dredged from the floor of the Trough are of Upper Cretaceous age, but there is every reason to believe that the majority of the basal infill is older. It is currently thought that the Trough may have first appeared as a downfaulted basin structure which acted as a collecting ground for coarse land-derived sediments from as early as the beginnings of the Mesozoic (Triassic period) or possible even earlier in the uppermost Palaeozoic (Permian period).

Following distension of the continental crust along the line of Porcupine and Rockall Trough during the early stages of rupture between the North American and European Plates, a long linear fault-bounded trough developed from the latitude of the Faeroe Islands in the North, southwards to the latitude of Southern Ireland. At that stage the Trough was very similar in structure to the present day East African rift valley and later to the submerged oil and gas bearing troughs of the North Sea rift system (see Section 2.4). The similarity in origin and sedimentary environment suggests that there is a likelihood of the prospective Permian and Triassic continental-type sequence of sediments, known from the North Sea troughs, being present near the base of the infilling succession in the Porcupine and Rockall Troughs.

From an analysis of the magnetic reversal pattern across the ocean floor of the North Atlantic, and the subdued magnetic features in Rockall Trough, it seems probable that spreading of the floor of the Trough may have occurred during Mid or Late Jurassic to Mid-Cretaceous times, ceasing before Upper Cretaceous time. Fossil pollen spores preserved in the pre-Middle Jurassic sediments of the small continental fragment of Orphan Knoll, which lies adjacent to the edge of the continental shelf West of Newfoundland, show very strong similarities to spores preserved in sediments of a similar age in northern England, suggesting that the continents were at that time continuous.

The break-up of the North American-Greenland Plate in the Upper Cretaceous (76 million years ago) and the rapid extension of the mid-Atlantic ridge system northwestwards into the Labrador Sea led to considerable local subsidence over the associated continental shelf regions, submerging parts of Rockall Plateau, and much of the present shelf area west of Britain. This subsidence may also correlate with the establishment of an Upper Cretaceous deep water marine environment throughout Rockall and Porcupine Troughs. Such a change is borne out by data from a JOIDES borehole, site 111, drilled on Orphan Knoll, which shows Upper Cretaceous marine sediments overlying a shallow water suite of limestones and carbonate sandstones. Prior to the opening of the Labrador Sea, the continental fragment of Orphan Knoll is believed to have lain adjacent to the mouth of Rockall Trough.

The Lower Tertiary separation of Rockall from southeast Greenland, was accompanied by the intrusion of hot granite rocks into the adjacent continental crust and the extrusion of volcanic lavas across the surface of Rockall Plateau, and the shelf area west of Britain.

This volcanic activity persisted into the Lower Eocene. Following an initially high rate of ocean floor spreading about the Reykjanes Ridge, the spreading rate began to slow down, producing a second phase of widespread subsidence throughout the

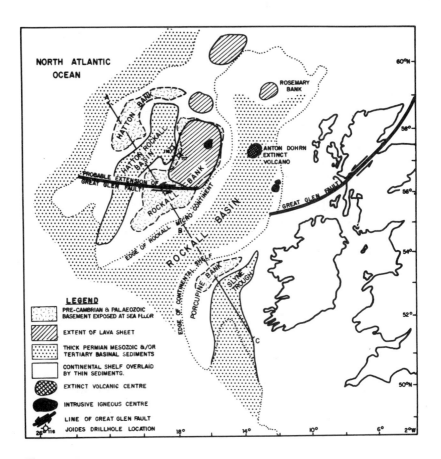

Figure 18
Geology of Rockall and Porcupine Banks.

Rockall Plateau forms a small fragment of continental crust, or microcontinent, in the North Atlantic, separated from the continental shelves of both Western Europe and Greenland. The nature of the northern extremity of Rockall Plateau is still uncertain, but it is thought that it may represent a rather broken continental link with the shelf west of the Shetland Islands. Although no accurate dating of the oceanic crust beneath Rockall Trough is available, it is believed that the Trough may have first developed during the Permo-Triassic period, as a downfaulted basinal structure, with continental rupture and the initiation of seafloor spreading occurring during the subsequent Cretaceous period. Due to the sparcity and lack of readily identifiable magnetic lineations across the floor of the Trough, it is now generally believed that sea-floor spreading spanned a period from late Jurassic times to mid-Cretaceous times, during a period of relatively few magnetic reversals. Note that closure of Rockall Trough brings in line the Great Glen Fault with the major fault present across Rockall Plateau.

A strong pattern of magnetic lineations to the west of Rockall Plateau permits an accurate lowermost Tertiary identification (60 million years ago) for the date of rupture and initiation of sea-floor spreading between Rockall and Greenland.

Rockall region during Upper Eocene and Lower Oligocene times. During this time Rockall Plateau became entirely submerged, with the possible exception of the tiny islet of Rockall itself, and a deeper water environment, persisting into the present day, became established throughout the Trough, leading to the deposition of very fine-grained sediments such as clays and oozes.

Rockall Plateau can conveniently be divided into three physiographic units as shown in Figure 18:

Hatton Bank — lying along the northwestern margin of the Plateau and having an overall NE-SW trend which swings to E-W at the northern end. Much of the bank lies between 1,000 and 1,500 feet below sea level.

Rockall Bank — lying along the southeast margin of the Plateau and culminating in Rockall islet. The Bank has a well defined NE-SW trend and water depths, although often very shallow, sometimes reach up to 1,300 feet.

Hatton-Rockall Basin — an intervening deep water (up to 4,000 ft) trough closed at its northern end by the George Bligh Bank and open at its southern end into the deep water of the North Atlantic Ocean.

Geophysical data suggests that although there is no evidence of oceanic crustal material on the floor of the Hatton-Rockall Basin, the underlying continental crust is considerably thinned, and it seems likely that one of the attempts at extending the North Atlantic spreading ridge between Greenland and Europe may have occurred along the line of this basin. Following cessation of movement along this line, in probably Mid to Late Cretaceous times, the trough became the collecting ground for up to 9,000 ft of continentally derived and marine sediments.

During part of the JOIDES programme of deep sea investigations, two drill holes (sites 117 and 116) were located on the margin and close to the central axis of Hatton-Rockall Trough respectively (see Figure 19), and as a result there is good evidence for the dating of much of the Upper Tertiary sediments within the Trough, although the age of the underlying horizons still remains speculative.

6.2   **Geophysical and Bottom Sampling Data**

Seismic refraction and reflection profiles across the Plateau show a thin veneer of sediment overlying economic basement on both the Hatton and Rockall Banks, while the central Hatton-Rockall Trough and Rockall Trough to the west display a thick infilling sequence of layered sediments interrupted by a number of prominent reflectors. The interpretative sections D and E, shown in Figure 20, are based on two actual seismic lines shot across the Troughs. Both lines form part of a commercial seismic programme run between Rockall Plateau and the shelf west of Britain, and have been included to show some of the typical seismic features of the basins.

Magnetic data show the Plateau to be very varied in character, by comparison with both the ocean floor of the North-East Atlantic to the west, where the magnetic *striping* is strongly marked, and with the weakly disturbed magnetic field of the sediment-filled Rockall Trough to the east. The appearance of a pattern of incoherent magnetic features broken by a few stronger correlatable lineations in the Trough indicates that much of the spreading may have occurred during a Lower Cretaceous quiet magnetic interval which is noted elsewhere in the North Atlantic for its few periods of magnetic reversal. The absence of any correlateable Tertiary magnetic anomalies within the Trough suggests that spreading had ceased by the end of the Mesozoic.

On Rockall Bank, the presence of a large Tertiary intrusive centre partially represented by Rockall Islet and confirmed by the dredging of basalt across the Bank suggests that many of the high amplitude magnetic values displayed across the Plateau are caused by extensive sheets of lava and volcanic debris. Dredging over the southern part of the Bank produced samples of highly deformed hard crystalline rock of Precambrian age similar to those in southern Greenland and western Scotland, and seismic profiles suggest that most of the Plateau is composed of this rock which is unprospective for oil and gas exploration.

6.3   **Stratigraphy**
Hatton-Rockall Basin
Up to 9,000ft of near-horizontal sediments infill the Hatton-Rockall Trough. These comprise an upper unconsolidated and semi-consolidated section of younger Upper and Mid-Tertiary rocks in the centre of the basin, and an older section of consolidated Lower Tertiary and Mesozoic rocks. The older rocks form a uniform, gently downwarped layer which appears to rest directly on the metamorphic basement floor of the Trough. Around the margins of the Trough, the older sediments are exposed on the sea-floor, but towards the centre of the basin they are consecutively overlain by onlapping younger rocks of the upper sedimentary section.

Two boreholes drilled in the Hatton-Rockall Trough during the JOIDES deep-sea investigation, showed that apart from two minor breaks, continuous deposition of calcareous oozes and

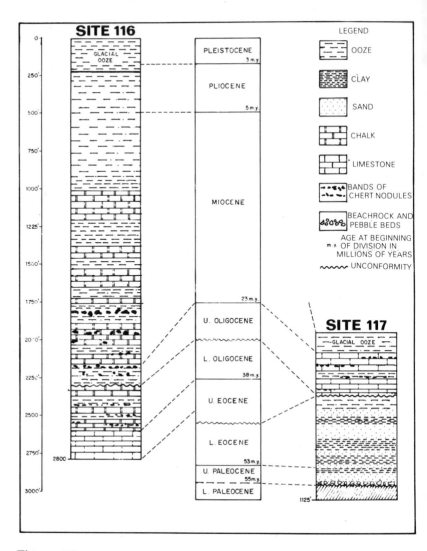

Figure 19

Stratigraphic sections penetrated by two deep water Joides boreholes on Rockall Plateau.
For borehole locations see Figure 18.

In an effort to provide a definitive test of some of the earlier concepts and predictions of sea-floor spreading, JOIDES (Joint Oceanographic Institutions Deep Earth Sampling) set up a major scientific project which entailed the drilling of a large number of deep-water boreholes throughout the oceans of the world. During the course of this project, in the summer of 1970, the JOIDES drill ship Glomar Challenger drilled a series of holes across the North Atlantic between Western Europe and Northeast Canada. Two of these holes were sited on Rockall Plateau to investigate the sediments infilling the Hatton-Rockall Basin and to learn something about the history of subsidence and sedimentation of the Plateau. Site 116 was drilled in 3715' of water in the centre of the basin, and Site 117 was drilled in 3354' of water on the western flank of Rockall Bank.

chalk has taken place within the Trough from the beginning of the Tertiary period until the present day.

With the slowing down, 47 million years ago, of ocean-floor spreading about the Reykjanes Ridge to the west, a period of widespread subsidence began to affect the entire Plateau, causing prolonged sinking of the floor of the Trough throughout the Upper Eocene and Lower Oligocene. Following this rapid change in the environment of deposition, and the marked reduction in potential sediment source areas, the sedimentation rate throughout the region became considerably reduced and for a time was zero. The result of this was that a new sequence of much deeper water deposits was laid down unconformably across the older pre-Eocene sediments in the centre of the Trough.

Renewed downwarping of the basin during the Oligocene brought a second break in sedimentation and introduced a new unconformity. With the stabilisation of spreading about the Reykjanes Ridge in Upper Tertiary times, active sedimentation was once more resumed across the Trough, and a thick sequence of deep-water deposits was laid down.

Seismic reflection profiles across the basin in the vicinity of JOIDES boreholes 116 and 117 can be correlated to show the top 45% of the infilling sedimentary section (approximately 2,500ft) is composed of soft semi-to unconsolidated Upper Tertiary, Miocene and Pliocene chalks and calcareous oozes, overlain by a thin veneer of Pleistocene glacial ooze (see Figure 19). Within the centre of the basin the Upper Tertiary beds overlie Oligocene and Lower Miocene cherty limestones, which themselves rest unconformably on a relatively thin sequence of well consolidated pre-Eocene sediments. The age and nature of the sediments underlying the Eocene unconformity throughout most of the basin, is somewhat conjectural. The only indication of their character comprises a section of Palaeocene graded clays and sands sampled at the base of drillhole site 117, close to the edge of the Trough. Because of the marginal situation of this sample, it may not be representative of the majority of the section which infills the centre of the Trough. However, the tectonic history of the Trough as a failed arm of crustal spreading suggests the possibility that much of the basal sequence is composed of coarse continentally derived sediments such as sandstones and conglomerates, laid down under shallow water or terrestrial conditions.

Rockall Trough

Seismic reflection profiles between Ireland and Rockall Plateau show that the present day topographical basin of Rockall Trough is only a relic of an originally much deeper basin feature now partially infilled by 10,000ft or more of roughly horizontally bedded strata. The greatest thickness of sediments within the

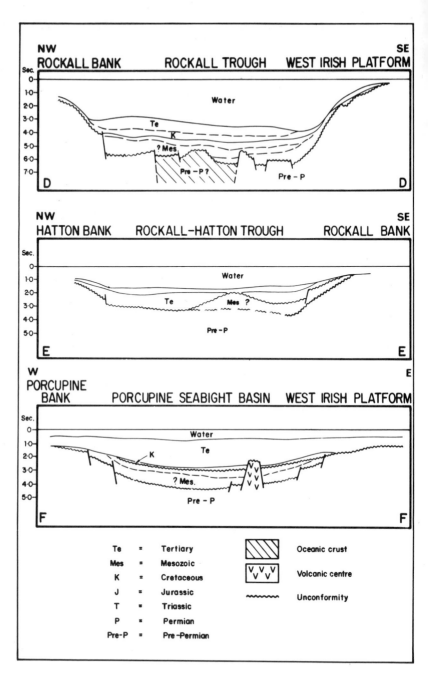

Figure 20
Sections based on seismic profiles across

D — D *Rockall Trough*
E — E *Rockall-Hatton Trough*
F — F *Porcupine Seabight Basin*

For locations of seismic profiles see Figure 11.

Trough occurs on the eastern side close to the edge of the Irish-Scottish continental shelf, where continuous deposition is thought to have taken place since the Trough was first created as a primitive rift structure.

Seismic reflection data across the Trough show that the infilling sequence of Tertiary and Mesozoic sediments contains three prominent reflecting horizons. These reflectors represent unconformities within the succession related to periods of rapid sinking of the trough floor and a reduced inflow of sediments. Due to the lack of geological control across Rockall Trough, the age of the infilling sediments and of these prominent reflectors is unknown and a probable age can only be assigned to them by reference to the geological history and structural development of the Trough, and its similarity in seismic appearance to the smaller Hatton-Rockall Trough on the West.

The oldest unconformity appears to be a result of considerable downwarping of the Trough in association with the Upper Cretaceous extension of the mid-oceanic active spreading ridge system into the Labrador Sea. Thus although the age of the basal sediments underlying the unconformity is uncertain, it seems likely that they will date back from Upper Cretaceous times to the time of formation of the Trough as an embryonic rifted structure in the Trias or Permian. Such sediments will have been laid down in terrestrial or shallow marine environments and as such will be characterised by irregular units of fluviatile and lacustrine sandstones, conglomerates and mudstones with the subsequent invasion of the sea across the Trough floor, the continental deposits will probably pass up into a layered sequence of shallow water carbonates and evaporites.

Following the first phase of widespread crustal subsidence in the northeast Atlantic, the floor of the Trough (in particular the southern part), sank and a regime of moderately deep water sedimentation was established. Elsewhere over southern Britain and the southern North Sea, the same subsidence is represented by a major break in Cretaceous sedimentation, and marks the sudden appearance of the extensive chalk-depositing seas. From the geophysical character and apparant thickness of the sedimentary interval between the lower and middle reflectors, it seems likely that this marine environment extended westwards across parts of Ireland and the Celtic Sea to deposit a similar Upper Cretaceous chalk-greensand-clay sequence over the Trough.

The intermediate unconformity in Rockall Trough is believed to correspond to the same Upper Eocene period of widespread subsidence and decreased rate of sedimentation identified in the Hatton-Rockall Trough to the west, and likewise appears as a fairly strong, continuous reflector across the entire basin. Despite the decrease in sedimentation rate, continued sedimentation over the axis of the basin occurs simultaneously with subsidence. Consequently the thickest and most complete sequence of post-Upper Eocene deposits is developed only within the centre of the Trough, while towards the margins older sediments overlie the unconformity. Lithologically these sediments are likely to be semi-consolidated limestones and calcareous oozes comparable to those encountered in the Hatton-Rockall Basin, although they may contain a higher proportion of coarser land-derived material such as sands and silts along the eastern margin of the Trough where it follows the edge of the continental shelf.

The uppermost unconformity indicates a second period of Tertiary subsidence within Rockall Trough related to a change in the pattern of sea-floor spreading in the North Atlantic during the Oligocene, 47 million years ago. A similarity in the seismic characteristics of the reflectors overlying this unconformity with those identified in the Hatton-Rockall Basin, suggests that the sequence may also be composed of Upper Tertiary unconsolidated oozes and semi-consolidated chalk.

Selected Reading

BAILEY, R J	1974	Seismic reflection profiles of the continental margin bordering the Rockall Trough. J Geol Soc London Vol 130 pt 1.
LAUGHTON, A S	1972	The southern Labrador Sea — a key to the Mesozoic and early Tertiary evolution of the North Atlantic in Laughton, A S et al 1972. Initial Reports of the Deep Sea Drilling Project XIII, Washington.
LAUGHTON, A S	1972	JOIDES deep boreholes. A discussion on the geology of Rockall Plateau. Proc Geol Soc.
ROBERTS, D G	1971	New geophysical evidence on the origins of the Rockall Plateau and Trough. Deep Sea Res Vol 18.

7   Porcupine Seabight

7.1   Introduction

Off the southwest coast of Ireland, the west Irish shelf takes a broad sweep westwards to form a submerged peninsula of continental crust isolated from the main continental shelf region to the east by a northward extending tongue of deep water, the Porcupine Seabight (see Figure 18). The peninsula, known as Porcupine Bank, forms an extensive shoal area in less than 1,500 feet of water and like much of the continental shelf around Britain including Rockall Plateau, is formed of ancient metamorphosed and folded rocks of Precambrian and Palaeozoic age overlaid by a thin veneer of Tertiary deposits. These in themselves provide little interest for hydrocarbon exploration.

Porcupine Seabight is the surface expression of an underlying deep northward-narrowing sedimentary trough now partially infilled by up to 5,000 feet of Mesozoic and Tertiary sediments. The trough appears to have formed as a major tensional feature. The underlying thinned continental crust, suggests that it is the result of one of several abortive attempts during the Mesozoic to extend the axis of mid-Atlantic sea-floor spreading northeastwards between Europe and Greenland. Apart from tensional faulting there is very little tectonic deformation of the trough and bank region and the floor of the trough appears to have been gently subsiding throughout Mesozoic and Tertiary times.

7.2   Geophysical Data

Because of the considerable depth of water infilling the trough (over 4,000ft) no dredge samples have yet been obtained of the upper surface of the infilling sedimentary succession, and the age and nature of the basin infill has to be determined entirely on the geophysical characteristics of those horizons which can be identified on seismic reflection profiles. Seismic reflection data shows a threefold subdivision of the infilling sedimentary series over most of the Seabight (as illustrated by Section F Figure 20). This comprises an upper series of undeformed and unconsolidated young strata resting on a similar intermediate sequence, which in turn overlies a lower unit of gently dipping rocks. Towards the northern end of the Seabight the intermediate sequence thins and eventually disappears so that the uppermost sequence rests directly on the basal folded sequence. The reflection data shows that at its southern end the Seabight has a typically trough-like profile with partly fault-controlled margins characteristic of rift systems associated with the early separation of two continental

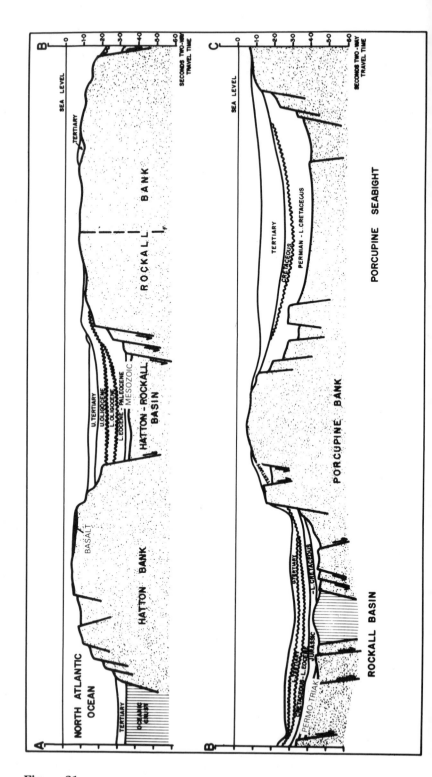

Figure 21
Diagrammatic geological section across Rockall Plateau and Porcupine Seabight. For section locations see Figure 18.

plates. At its northernmost end the feature continues across the shoal region of Porcupine Bank as a markedly shallower, narrow, sediment-filled trough, the Slyne Trough. Unlike the Seabight Trough, this Slyne Trough has no expression in the present-day bathymetry.

7.3   Stratigraphy

In the absence of drillhole data, and therefore any form of geological control over the subsurface of the Seabight Trough, interpretation of the possible ages of the three sedimentary intervals present, and the period of non-deposition separating them, can only be related to characteristically similar seismic sequences known to the northwest and west in Rockall Trough, and also to the southeast in the Celtic Sea troughs where some geological control is available.

The lowest unit comprises a sequence up to 4,500 feet thick overlying the uneven block-faulted floor of the trough. These sediments are considered to have been deposited during the Permian, Triassic and possibly Jurassic periods under continental and shallow marine conditions, when the basin was at an early rift stage of continental separation. Sedimentation at that time is likely to have been dominated by the accumulation of coarse land-derived material, such as sands, gravels and conglomerates eroded from the steep margins and deposited across the floor of the trough.

With the continued sag of the trough floor, the development of lagoonal and subsequently shallow-water marine conditions probably led to both a change in the grain-size and chemical composition of the sediments being laid down. The earlier predominantly sandstone regime probably gave way to a sequence of carbonates, silts, clays and evaporites. Near the northern end of the Seabight the appearance of strong dome-like disturbances in the uppermost sedimentary series is suggestive of intrusive salt features which originate from the underlying basal sequence, and lends confirmation to the possible existence of extensive salt horizons of Triassic or Lower Jurassic age within this unit.

Geophysical evidence suggests that the lowest unit in the Slyne Trough is a northward continuation of the basal unit of Triassic to Middle Jurassic continental and shallow-marine sediments thought to be present in the Seabight Trough.

The intermediate unit which is absent at the head of the Seabight, extends seawards as a southwestward thickening wedge between the eroded surface of the gently dipping Jurassic horizon and the overlying near-horizontal bedded Tertiary sediments.

From a marked similarity in seismic character to the sequence of sediments developed on the edge of the continental shelf to the northwest which borders Rockall Trough, it seems likely that these sediments are chalks of Late Cretaceous to Lower Tertiary (Palaeocene) age. The base of this intermediate unit is believed to represent the sudden advance of the chalk seas into the Seabight, which followed a period of widespread subsidence of much of the continental shelf area west of Britain in the Mid-Cretaceous. The Lower Cretaceous is thought to be missing over much of the trough.

The overlying younger strata form a thick sequence of uniformly layered Middle and Upper Tertiary sediments conformable in attitude to the marginal slope and sea-floor of the trough. The sediments extend northwards overstepping the gently dipping strata at the head of the Seabight to follow northeastwards into Slyne Trough. It is believed that the Tertiary, as elsewhere in the area, will consist of Miocene and Pliocene unconsolidated silty clays underlain by a Palaeocene-Eocene section containing a high proportion of chalks. A thin veneer of Pleistocene glacial oozes overlies the entire infilling section.

Selected Reading

CLARKE, R H et al   1970   Seismic Reflection Profiles of the Continental Margin West of Ireland.

SCOR Cambridge Symposium.
Vol 2 IGS Rep No 70/14.

8 Irish Sea Basins

8.1 Introduction

A series of deep sedimentary basins extends northwards from the Celtic Sea through the Irish Sea, northeastern Ireland and the Firth of Clyde to link up with the basins off western Scotland. In the Irish Sea these basins are dominated in part by the trend of the ancient Caledonian (Lower Palaeozoic) mountain chains still forming the main mountain ranges of Wales to the east and the Scottish Highlands to the north, and in part by a series of NNE-SSW fractures believed to be associated with the break-up of the European-North American continent. As a result the Upper Palaeozoic troughs off the east coast of Ireland and west of the Lake District and the Mesozoic-Tertiary troughs beneath Cardigan and Tremadoc Bay all show a dominant NE-SW basinal trend, superimposed in places by a weak, roughly North-South pattern of marginal faults.

Geological and geophysical evidence over the sea-floor between Wales and Ireland show that the southern part of the Irish Sea is separated into two distinct regions by a central upfolded basement ridge trending southwestwards from Llewn Peninsula in North Wales towards Wexford on the southeastern tip of Ireland. To the west of this ridge, the basement surface of the sea-floor is broken by a complex, downfolded basin, the *Central Irish Sea Basin*; while to the east, a pair of sub-parallel Mesozoic-Tertiary basins underlie the St George's Channel and the Cardigan-Tremadoc Bays. The northern part of the Irish Sea, between the north coast of Wales and the south coast of Kirkcudbrightshire is dominated by two major sedimentary basins, the *Solway Firth Basin* and the *Manx-Furness Basin* (see Figure 22).

The *Central Irish Sea* Basin underlies the western half of the Irish Sea. At its northern end the trough bifurcates to form an eastern arm beneath *Caernarvon Bay* and a northern arm to the west of Anglesey. Gravity and seismic data across the main basin suggest that the major infill is composed of some 10,000ft of Carboniferous and younger sediments similar to, and forming an offshore extension of, the Carboniferous limestone-calcareous mudstone sequences in southeastern Ireland, and likewise, through the early destruction of primary porosity in the limestones, may also be conspicuously lacking in the basic source rock and reservoir properties needed for hydrocarbon generation and accumulation.

Beneath Caernarvon Bay, the eastward arm forms a broad NE-SW trending downfolded basin, sharply downfaulted on one side

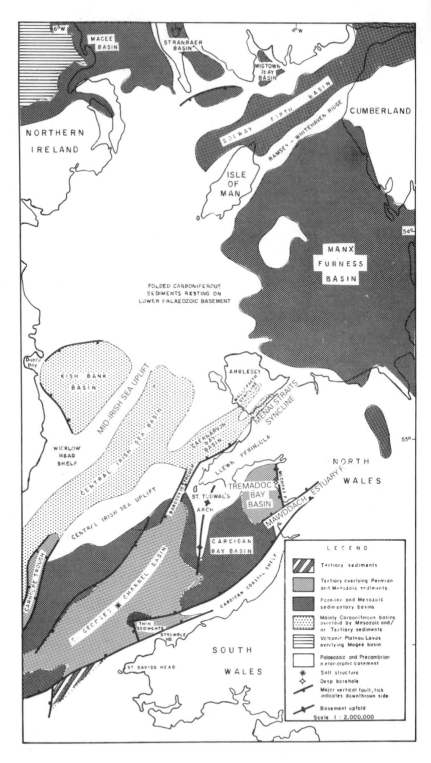

Figure 22
Geological Map of the Irish Sea.

against the basement ridge of the Llewn Peninsula and on the other gently onlapping across the basement high of the Holy Island Shelf. Geophysical data show the presence of a deep fault controlled trough, the *Barnsey Island Trough* (roughly 2 miles wide) off the end of the Llewn Peninsula cross-cutting the Central Irish Sea basement ridge to link the *Caernarvon Bay Basin*, with the extensive *St George's Channel* (Permian-Mesozoic) Basin to the south. At its northeastward end, the basin shelves gently onto the ancient basement platform of Anglesey, continuing onshore as a pair of minor downfolded troughs, the Malltraeth Syncline and the Menai Straits Syncline. Both these basins are of interest as they provide a clue to the age and type of sediments likely to be present in the deeper basin offshore. In the Malltraeth Syncline almost 5,000ft of Lower Carboniferous limestones and mudstones, and Upper Carboniferous sandstones and coal measures are overlaid by a succession of pebble beds, sandstones and shales, known as the *Red Measures*, while further east in the Menai Straits Syncline the entire Upper Carboniferous section is absent, presumably having been removed by erosion, and a *Red Measure* sequence rests directly on 1,000ft of Lower Carboniferous limestone.

To the east of the central basement ridge, a pair of subparallel, deep Mesozoic-Tertiary basins underlie St George's Channel and the Cardigan Tremadoc Bays (Figure 22). Geophysical data show that the overall form of the St George's Channel Basin is that of a broad, downfolded and partially fault-bounded trough which roughly coincides with the main axis of the Channel. To the northwest and southeast the margins of the basin are flanked by the Caledonain trend of the *Centrcl Irish Sea Ridge* and the basement shelf off the north coast of Pembrokeshire respectively, while at the southernmost end the thick sedimentary infill extends southwestwards through a narrow channel into the broad sedimentary basin of the Northern Celtic Sea Trough. The structure of the central portion of the basin comprises two basement depressions separated by a central slightly shallower WNW-ESE trending saddle lying west of Strumble Head in Wales. Reflection profiling shows that the major infill of the basin is composed of 15,000—20,000ft of sediment downfolded into a broadly synclinal feature (Figure 23A). Although the syncline has an overall Caledonian trend, the axis of the fold follows a sinuous course with frequent local deviations from the main NE-SW direction.

Lying immediately offshore from west Wales, where only ancient metamorphosed rocks are exposed, a very thick sequence of Mesozoic and Tertiary sediments infill a northeastward plunging basinal structure beneath Cardigan Bay and Tremadoc Bay. To the northwest the basin is bounded by the Palaeozoic basement of Llewn Peninsula and St Tudwal's Arch, and similarly to the east and southeast by the sharply faulted basement margin of west Wales and the offshore Cardigan coastal shelf. To the south-west a shallow basement ridge, St Tudwal's Arch, now separates the Cardigan Bay Basin from the structurally similar St George's Channel Basin (Figure 22), although for much of their Mesozoic sedimentary history it is probable that they formed one continuous basin.

Two boreholes were sited at Mochas and Tonfanau on the east coast of Tremadoc Bay where the thick offshore sedimentary basin terminates abruptly against faulted basement margin of onshore Wales, in an effort to determine the geological succession and character of the marginal sequence of the Tremadoc Bay basin terminates abruptly against the basement margin of lying Mesozoic rocks and confirmed the sequence previously suggested by geophysical and sea-floor sample data for the off-shore area. These boreholes also showed that the eastern margin of the basin is controlled by a major fault, the Mochras Fault, which downthrows the basement rocks of Wales by 15,000ft to the west beneath the younger sediments of Tremadoc Bay. This fault closely follows the coastline as far south as Mawddach Estuary where it is displaced approximately 3.5 miles westwards by a second NE-SW trending fault following the line of the Mawddach River. South of this, the margin of the Basin underlying Cardi-gan Bay becomes more complex, and is controlled by a number of interacting faults.

Lying immediately south of the Kirkcudbrightshire coast, in the Northern Irish Sea, gravity data show the Solway Firth to be underlain by a narrow, linear, NE-SW trending basin which extends from the northern tip of the Isle of Man towards the onshore Permo-Triassic basin of the Carlisle-Vale of Eden area. This is known as the *Solway Firth Basin*. Permo-Triassic rocks representing the margin of the basin are also exposed on the Isle of Man, and an interpretation of the geophysical data suggests these sediments thicken offshore towards the centre of the basin.

Figure 22 shows the Manx-Furness Basin to be a much broader feature by comparison with the other basins of the Irish Sea. Lying between the Solway Firth Basin and the north coast of Wales, it extends eastward from the Isle of Man towards the Lancashire coast, where it is contiguous with the coastal strip of Permo-Triassic rocks north of Liverpool. Along its northern margin, the basin is separated from the Solway Firth Basin by a ridge of older (possibly Carboniferous), rocks which stretch from Ramsey on the Isle of Man to Whitehaven in Cumberland. To the southeast exposures of Permo-Triassic rocks in the Liverpool-Wirral area show the basin to be connected to the deep, fault-bounded *Cheshire Basin* onshore.

8.2 Geophysical and Bottom Sampling Data

The first indications that the Irish Sea was underlaid by a number of very thick young sedimentary basins (which are either absent or in general poorly exposed onshore around the margin of the sea) became apparent after a regional gravity survey had been carried out across the area. This survey showed the Irish Sea to be dominated by generally high gravity values in comparison with the surrounding land areas, superimposed across which are a series of local anomalies. These correspond to the shape of the underlying basement structure, with the regions of relatively low gravity values correlating with the areas underlaid by sedimentary basins, and regions of relatively high gravity values correlating with the uplifted basement features.

Nine significant low gravity anomalies can be distinguished on the gravity data: an intense elongate low over St George's Channel, extending northeastwards as a subsidiary feature through Cardigan Bay into the closed gravity low of the Tremadoc and Barmouth Bay area; a similar slightly less intense feature over the western Irish Sea extending southwestwards from Caernarvon Bay towards Wexford; a minor low trending into the Menai Straits; a narrow tongue of low gravity linking the main St George's Channel-Cardigan Bay low with the Caernarvon Bay low; a complex low over the northeastern Irish Sea between the Isle of Man and the south Cumberland and Lancashire coasts; an oval shaped low extending from the Carlisle Basin across the Solway Firth; a narrow northwestward trending low through Luce Bay cross-cutting the Stranraer neck of land to join the Firth of Clyde low; and to the west a rectangular low overlying Kish Bank off Dublin Bay.

Subsequently over the southern Irish Sea a programme of seismic reflection profiling confirmed the broad findings of the gravity survey, establishing the depth and limits of the basins more accurately, and determining the age and nature of the sedimentary fill. Controls on the age and anticipated lithology of the individual horizons picked out by the reflection profiles across the sedimentary basins are provided by the deep boreholes of Mochras and Tonfanau which intersect the Tremadoc Bay basin margin encountering Tertiary sediments overlying a thick Lower Jurassic section and Triassic sandstones, and by a series of short core samples collected from the sea bed of Cardigan Bay and the southeastern half of the Irish Sea. Most of these samples have yielded sediments of Permo-Triassic, Lower-Middle Jurassic and Tertiary age. Seismic profiling west of the Welsh coast identifies three major seismic reflectors and four velocity groups within the sedimentary infill of the St George's Channel Basin. In the subsidiary Cardigan Bay Basin velocity layer 3 and seismic reflector 3 are missing. These are listed on the table below in conjunction with their probable age correlation.

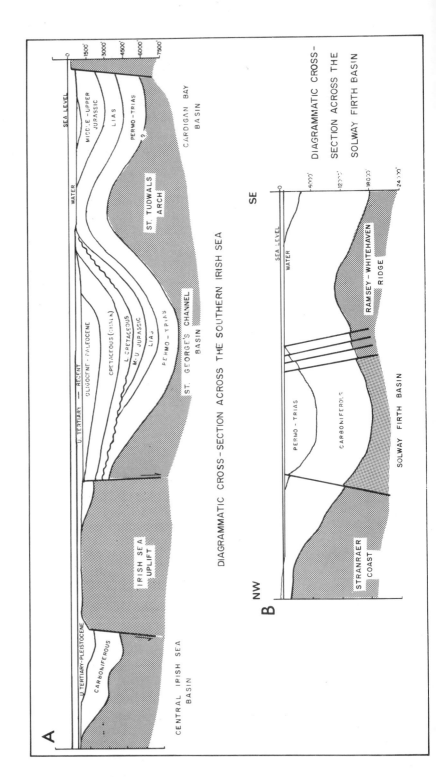

Figure 23

Diagrammatic cross-sections across

A   *The Southern Irish Sea*
B   *The Solway Firth Basin*

HORIZON DETECTED	MEAN VELOCITY	PROBABLE AGE CORRELATION
Sea floor		
Velocity Layer 1 **Seismic Reflector 1**	1.8 ft/sec	Upper Tertiary and Recent Miocene unconformity
Velocity Layer 2 **Seismic Reflector 2**	2.3 ft/sec	Lower Tertiary Top Cretaceous Chalk unconformity
Velocity Layer 3 **Seismic Reflector 3**	3.5 ft/sec	Upper Cretaceous Chalk Base Cretaceous Chalk
Velocity Layer 4 **Seismic Reflector 4**	4.9 ft/sec	Permo-Triassic and Jurassic Palaeozoic metamorphic Basement

Beneath the western part of the Irish Sea the *Central Irish Sea Basin* is defined by an elongate Caledonian trending gravity low bounded to the northwest and southeast by the Mid-Irish Sea Uplift and the dividing barrier of the central basement ridge and Holy Island Shelf respectively. At its northern end the basin is linked into the *Caernarvon Bay Basin* across a submerged basement sill. Seismic profiling across both the Central Irish Sea Basin and Caernarvon Bay Basin show a twofold grouping of the sedimentary infill, comprising a lower series characterised by laterally persistent thick-bedded strong reflectors, and a southward thinning upper series characterised by impersistent, thin-bedded weak reflectors. The two groups are separated by a slight angular unconformity. Following deposition of the lower group, a period of movement compressed the basinal sediments into a series of gently folded structures running roughly along the length of the trough and as such has markedly affected the distribution and thickness of both this and the upper group of rocks. However, the axes of these folds lies at a slight angle to the Caledonian trend of the main basin and this has given rise to a rather complex structural picture, particularly over the southern half of the basin. Gravity and seismic data show that the margins of the northern end of the basin and the Caernarvon Bay trough are controlled by steeply dipping faults with vertical displacements in the region of 15,000ft, but towards the south the sediments tend to thin towards the edge of the basin, which in addition to their more complexly folded nature makes the limits of the basin harder to define.

As there is no direct borehole or sea-floor sample evidence for the fill of the Central Irish Sea Basin, the character of the sediments is largely determined by analysis of the seismic reflection profiles and by correlation with coastal exposures in south-east Anglesey and Wexford. In addition, velocity values obtained from a single seismic refraction station located over the centre of the main basin, show the development of a two-fold velocity sequence, comprising an upper low velocity layer some 2,000ft thick, and an underlaying high velocity layer. Values in the order of 1.8 — 2.1ft/sec obtained from the upper layer, suggest an Upper Mesozoic or younger age for the sediments, while those from the underlying layer correspond to values normally associated with rocks ranging in age from Lower Paleozoic to Mesozoic. The southward thinning of the upper group is confirmed by this refraction station, and towards the southwestern end of the basin the pattern of folds in the lower series can be traced onshore to correlate with the structure of the Carboniferous rocks around Wexford. Over the northern part of the basin, and beneath Caernarvon Bay, both groups are thickly developed. In general, the lower series reaches thicknesses in the order of 2,000 — 3,000ft within the central downfolded core of the basin; while the upper series shows a more variable thickness as a result of being deposited after the main period of folding had ceased, and reaches values of 4,000ft along the axis of the downfolded structures, decreasing to less than 1,500ft over the upfolded regions. At the northern end of the Caernarvon Bay Basin, Carboniferous rocks ranging in age from Lower Palaeozoic to Mesozoic. The trending North-East from Malltraeth and the other follwing the line of the Menai Straits. Over the Malltraeth Coalfield the *Red Measures* sequence of sandstones, mudstones and pebble beds appears to thicken southeastwards, indicating that a similar Carboniferous sequence may exist offshore beneath Caernarvon Bay. Seismic profiles across the bay suggest that the upper seismic group containing a sequence of Permo-Triassic and younger rocks thickens progressively southwestwards across the Caernarvon Bay Basin and into the main Central Irish Sea Basin.

Immediately to the east of Dublin, the *Kish Bank Basin* is defined by a rectangular gravity low some 35 miles long and 25 miles wide. The anomaly is characterised by very steep gravity gradients along both the northern and southwestern edge the basin, indicative of fault-bounded basin margins, while to the east it shows a gentle gradient onto the Mid-Irish Sea basement uplift. Seismic reflection coverage of this basin is limited.

Over the northeastern part of the Irish Sea, shallow seismic reflection data suggest the presence of flat-lying and gently folded beds within the *Manx-Furness Basin*, which appear to be a continuation of the Permo-Triassic coastal outcrops of Lancashire and Cumberland. Shallow coring of the sea bed has established the presence of red Triassic mudstones, while onshore boreholes close to the Lancashire coast have encountered several thousand feet of Permo-Triassic rocks.

Further north, shallow seismic reflection data have also confirmed the presence of a sediment-filled trough underlying the

Solway Firth (see Figure 22). Magnetic and seismic data suggest that the margins of this basin may be fault controlled. The seismic profiles indicate a two-fold sequence of sediments infilling the *Solway Firth Basin*, comprising a thick lower group characterised by poor seismic reflectors, which may represent Permo-Triassic strata, and a thin upper group, restricted to the central axis of the basin of probable Lower Jurassic rocks. Following the northern margin of the basin is a narrow belt of structural disturbances which are believed to indicate the presence of salt within the basin.

Oil company interests have so far been concentrated over the northeastern portion of the Irish Sea, where following the shooting of a number of commercial seismic lines west of the Lancashire-Cumberland coast, two deep wells were drilled in the Manx-Furness Basin by a Gulf Oil-National Coal Board consortium. The boreholes went to depths of 8,500 ft and 10,400 ft respectively, and although neither hole encountered commercial hydrocarbons, the sequence penetrated is believed to have been largely of Permian and Triassic age.

8.3 **Stratigraphy**

Irish Sea Basins

Direct information on the nature of the sediments infilling the Kish Bank, Central Irish Sea, Caernarvon Bay and St George's Channel Basins has proved impossible to obtain by shallow gravity coring of the sea-floor, since much of the anticipated Upper Palaeozoic, Mesozoic and Lower Tertiary basinal sequence is obscured beneath a thin but persistent cover of young Tertiary and recent sediments. Thus, until deep marine drilling takes place, any interpretation of the infill is dependent upon the analysis of the available geophysical records across the basins, in conjunction with apparent correlatable seismic horizons in the Cardigan-Tremadoc Bay Basin, where knowledge of the infilling sedimentary sequence is based on borehold data along the Welsh Coast and offshore in the bays.

A series of seismic profiles across the centre of Cardigan Bay show a basinal succession of downfolded Permo-Triassic sediments, overlain by a central core of Lower and Middle Jurassic rocks. Along the northwestern margin of the basin, Permo-Triassic sandstones, calcareous mudstones and salt layers in the *Keuper facies* form a broad band some three miles wide, exposed on the sea-floor. These beds rest directly on the ancient basement of St Tudwal's Arch and appear to be the oldest sediments preserved in the basin. Along the complementary southeastern margin of the basin the Permo-Triassic rocks terminate abruptly against the fault-bounded Palaeozoic basement forming the Cardigan Bay shelf, and as a result are concealed beneath a similarly truncated sequence of younger Mesozoic sediments.

Lower Liassic (basal Jurassic) mudstones and siltstones, conformably overlying the Permo-Triassic, form a parallel band in excess of five miles wide along the northwestern margin of the basin, but are similarly concealed along its south-eastern margin. Middle and Upper Liassic rocks are as yet unknown from sea-floor sample data within the basin, and reflection profiles indicate the presence of a small angular unconformity near the central axis of the basin which suggests that Middle Jurassic rocks rest directly on the eroded surface of the Lower Liassic beds. The seismic character of the Lower Lias in the outer part of Cardigan Bay shows few and well-spaced intra-Liassic reflectors, similar to the appearance of the Lower Liassic mudstones and siltstones in the Mochras borehole.

Middle Jurassic rocks form the major infill to the core of the basin and seismic reflection data suggest that they may reach a maximum thickness in excess of 4,000ft along the central axis. From their seismic character the lithologies of the Middle Jurassic rocks are in sharp contrast to those of the Lower Jurassic and are likely to be dominated by a carbonate facies similar to the Jurassic limestones of the Cotswold Hills to the southeast. Confirmation of this is suggested by offshore sea-floor sample data, and by the discovery of large blocks of Jurassic limestone, thought to have been originally eroded from the floor of the basin by ice-sheet action, which are now interbedded in glacial clay along the Cardigan Bay coast near Tonfanau. Seismic profiles show the Middle Jurassic rocks to be characterised by a rhythmic sequence of evenly spaced strong reflective horizons indicative of a basic unit of deposition approximately 60ft thick. Upper Jurassic strata may also be present within the centre of the basin but due to their probable similarity in seismic character they have not been indentified to date from either offshore borehole or seismic profile evidence. During the following Cretaceous period, deposition appears to have been entirely absent over the bays.

In the northern part of Cardigan Bay and in Tremadoc Bay, seismic profiles show a substantial thickness of Tertiary and Quaternary sediments overlying the downfolded Permo-Triassic and Jurassic sediments in the basin, and extending across the basin margins to rest directly upon the Palaeozoic basement rocks. More than 200ft of marginal Tertiary sediments have been cored in the coastal borehole at Tonfanau, where they appear to rest on a steep westward dipping fault surface of Lower Palaeozoic basement. Similarly in the Mochras borehole, over 1,800ft of Tertiary carbonaceous clays and red pebble beds resting unconformably on Lower Jurassic rocks, have been encountered. However, offshore over the main basin, seismic data suggest that the average thickness of the Tertiary sediments may be in the order of 1,200ft.

A thick mantle of unconsolidated Quaternary deposits covers the entire Tremadoc Bay sea-floor, with the exception of small areas on the St Tudwal's Arch, where basement rocks are exposed. The thickness of this veneer is difficult to determine on seismic profiles, but sea-floor borehole data indicates typical thicknesses of about 300 ft.

Deep seismic reflection data across the adjoining St George's Channel Basin to the west (of which the Cardigan Bay Basin formed a subsidiary arm throughout much of the Permian and Mesozoic depositional history) show that the margins of the basin are dominated by a series of NNE trending, near-vertical faults, successively downfaulting basement rocks beneath the central axis of the basin. As in the Cardigan Bay Basin, the St George's Channel Basin appears to contain a thick, downfaulted and folded basal succession of Permo-Triassic rocks with a central core of downfolded younger Mesozoic and Tertiary rocks across which are overlain a series of flat-lying unconsolidated young sediments of possible Upper Tertiary (Neogene) and Pleistocene age. Along part of the eastern edge of the basin, off the Pembrokeshire coast, a thick succession of Mesozoic and Tertiary sediments overlie the downfaulted basin margins, onlapping eastwards across the Lower Palaeozoic basement of the Welsh offshore shelf. Further south, off St David's Head, the block-faulted margin terminates abruptly against a NNW trending fault, which downthrows basement rocks to the south beneath a thick succession of Mesozoic sediments.

To date there is no direct evidence for Permian and Lower Triassic sediments forming the lowest strata in the basin, although seismic records indicate their likely existence. Upper Triassic, reddish-brown, gypsiferous mudstones and siltstones, of presumed 'Keuper' facies, are known from sea-floor borehole data on the eastern margin of the basin. To the west of Strumble Head, Pembrokeshire, strong local disturbances, indicative of salt tectonics, are apparent on the seismic reflection profiles across the area, and suggest that the Upper Triassic mudstones, siltstones and evaporites may be accompanied by a thick but inconsistent horizon of salt in some parts of the basin. However, as the seismic records do not indicate the approximate horizon from which the disturbances originate, the salt could be of Permian and not Triassic age.

Sea-floor drilling on the flanks of St Tudwal's basement arch confirms the westerly extension of the Lower Jurassic (Lias) mudstones and siltstones from Cardigan Bay Basin into St George's Channel Basin. Liassic sediments have also been encountered in part of the step-faulted complex forming the southeastern margin of the basin. Similarly, reflection data suggests that the thick Middle and possibly Upper Jurassic section, known in Cardigan Bay, may extend across part of St Tudwal's Arch into St George's Channel Basin; but as yet there is no direct confirmation of their presence. No Cretaceous sediments have been encountered in boreholes within the southern Irish Sea but, although absent to the west in Cardigan Bay, they form a major part of the infill in the Celtic Sea basins to the south. In addition, seismic reflection profiles strongly suggest their presence in at least the southern part, and possibly the deeper central region of the northern part, of the St George's Channel Basin.

Over the centre of the Basin, presumed chalk rocks of Upper Cretaceous age rest conformably on the gently downfolded older strata, but towards the western margin of the basin the sediments begin to develop a distinct angular unconformity, overstepping the eroded surface of successively older beds towards the flanks of the Central Irish Sea basement ridge.

Although there is no direct evidence for the identity of the sediments involved, the seismic velocity for both the beds above and below this unconformity are indicative of a Mesozoic age. Elsewhere over southern Britain, the Celtic Sea, English Channel and southern North Sea, a distinctive and widespread unconformity marks the beginning of submergence of these land areas beneath a growing Upper Cretaceous chalk sea, and it is likely that this same unconformity may have extended sufficiently far west to affect the St George's Channel region. The reflective character and apparent thickness of the presumed Upper Cretaceous strata overlying the unconformity, are suggestive of a thick chalk sequence, such as is developed onshore in southern England, underlain by a basal horizon of greensand and clay. The thickness of the Upper Cretaceous beds in the centre of the basin is in the order of 3,000 ft, with the Chalk accounting for the top three-quarters of the section.

Over most of the basin, Lower Tertiary (Palaeogene) sediments rest conformably on the underlying Cretaceous strata, but towards the basin margin a slight angular unconformity develops suggesting a break in sedimentation or phase of slight downwarping of the basin during the latest Cretaceous to earliest Tertiary period. The Palaeogene strata are characterised by a number of strong, laterally continuous internal reflectors suggestive of well-bedded marine sandstones and shales, unlike the non-marine coal-bearing clay-pebble bed sequence encountered to the east in the coastal borehold at Mochras and in Tremadoc Bay. The top of the Palaeogene is defined by a well marked erosion surface, which is thought to coincide with a phase of slight uplift during the Miocene, related to the Alpine mountain building movements in Western Europe. This unconformity truncates the surface of the underlying folded Palaeogene and Mesozoic structures and is overlain by flat-bedded younger Upper Tertiary (Neogene) strata.

Seismic profiles show this upper Neogene layer to be composed of two units, a lower massive section and an upper bedded section. The general lack of distinct internal reflectors in the lower section, coupled with its low apparent seismic velocity, indicate that it may be similar in lithology to the non-marine clays and pebbles forming part of the Tertiary section in Tremadoc Bay to the northwest, while the upper bedded section is characterised by a number of regular, laterally continuous reflectors suggestive of an unconsolidated marine clay-sand sequence. The thickness of the Neogene varies considerably across the basin, but reaches a maximum of 500 - 600 ft along the central axis.

West of the Central Irish Sea (basement) Uplift, no direct geological data is available for the infill of the *Central Irish Sea Basin* and its subsidiary basin beneath *Caernarvon Bay*. However, limited geophysical data across the area, in conjunction with a fairly detailed knowledge of the stratigraphy and structure of Carboniferous exposures on Anglesey and along the Wexford coast, suggest that both basins are dominantly infilled by a weakly metamorphosed lower sequence of folded Carboniferous limestones and calcareous mudstones roughly 2,000 - 3,000 ft thick, overlaid by a younger sequence of suggested Permo-Triassic to Jurassic age. These younger sediments, although very variable in thickness, reach a maximum of 4,000 ft in the central downfolded core of the main Central Irish Sea Basin but thin both to the north towards Caernarvon Bay and to the south towards Wexford in Ireland. Lithologically the rocks exposed on Anglesey comprise a more complete sequence of Carboniferous rocks by comparison with those developed in Wexford, and include an upper series of sandstones and coal measures overlying the more typical limestone-mudstone succession. Although this Lower Carboniferous unit is of little interest for hydrocarbon exploration, the likelihood of the Anglesey coal measures extending southwestwards beneath Caernarvon Bay and even into the northern part of the Central Irish Sea Basin, presents a very real possibility for the existence of a rich hydrocarbon source rock beneath the overlying younger unit.

Near the southern end of the Central Irish Sea Basin a narrow fault-bounded trough, the *Carnsore Trough*, cuts across the uplifted central basement ridge and links the post-Carboniferous sediments of this basin with the Tertiary basin of the Northern Celtic Sea. Seismic profiles east of County Wexford show the trough to be infilled by gently westward dipping probable Neogene sediments unconformably overlying a downfolded sequence of thick (1,5000 ft) Palaeogene and Mesozoic sediments.

In the northernmost part of the Irish Sea, no direct sea bottom sampling data exists for the identification of the sediments infilling the *Solway Firth Basin*. However, the northeastern end of the basin extends onshore into the Carlisle and Vale of Eden (Permo-Triassic) Basin, and a brief consideration of the rock succession exposed here may well be pertinent to an understanding of the sequence to be expected offshore.

The lowest sediments rest directly on the eroded surface of the ancient Palaeozoic basement rocks, and are composed of coarse sandstones and pebble beds of Permian age derived by rapid erosion of the surrounding uplands. These are overlain by a thick (up to 1,500 ft) red dune sandstone, the Penrith Sandstone. The presence of reptilian footprints in horizons within the sandstone unit, suggests the periodic existence of lakes and pools during this continental, semi-arid period of deposition. Much finer rocks, consisting mainly of dull red mudstones (the Eden Shales) overlie the Penrith Sandstone, and are themselves overlain by a further succession of sandstones.

Although much of this Permian sequence appears to have been deposited under terrestrial, desert-like conditions, there is some evidence of marine influence. At several levels, beds of gypsum and anhydrite, and occasionally thin dolomites containing marine fossils, interfinger with the dune sandstones and shales, suggesting that the onshore part of the Solway Firth Basin may have suffered infrequent flooding by the sea from the west. Further evidence suggests that for limited periods of time, a sea connection may have existed to the east across the Pennines into the Zechstein Sea (see Figure 29). There is little doubt that the Eden Shales are Late Permian in age. The overlying thick sandstone sequences are unfossiliferous but are assumed to be chiefly of Triassic age. Low gravity values over the Vale of Eden suggest that at least 1 km of Permo-Triassic rocks are present northeast of of Penrith.

West of Carlisle is a poorly exposed area of Lower Jurassic rocks comprising a sequence of dark shales with thin limestone bands. These suggest deposition under marine conditions at a time when the Jurassic seas made large incursions onto the old Permo-Triassic landscape. In summary, it appears that the Carlisle-Vale of Eden area is the landward extension of the present-day Solway Firth Basin, as it was during Permo-Triassic times. It is probable therefore, that the offshore area contains thicker sequences of marine rocks than are evident onshore.

Although shallow sea-bed coring between the Isle of Man and the mainland coast provide some indication of the lithology of the Permo-Triassic sediments infilling the *Manx-Furness Basin*, the exposures of a large part of the basin margin along the Cumberland and Lancashire coast from Furness to St Bees Head give a major insight into not only the lithology but also the stratigraphy of the offshore basin. Basal Permian scree deposits and coarse pebble beds rest directly against the high ground of the

Palaeozoic Lake District massif. Seawards, these coarse deposits are overlain by younger shales and limestones, broken by occasional intervals of anhydrite. This suggests a westward transition from a terrestrial to a marine environment of deposition (St Bees Evaporites and Shales). Fossils obtained from the limestone horizons suggest a Late Permian age, although probably somewhat older that the the similar limestone beds in the Vale of Eden. As in the Vale of Eden-Carlisle area, these pass upwards into a thick Triassic sandstone sequence (4,000 ft in places) and then into red calcareous mudstones of Upper Triassic age. The precise division between the Permian and Triassic strata cannot be identified in the Irish Sea basins because of the lack of fossil evidence in the red desert sediments. As in the case of the Solway Firth Basin, the offshore succession infilling the Manx-Furness Basin is probably thicker and more marine in character than that of the onshore basin margin.

Cheshire Basin

Although this book is concerned chiefly with the offshore areas west of Britain, there are major differences between the overall geology of the land and sea areas and the coastline often forms a very imprecise boundary between the two provinces. This is no where better demonstrated than in the landward extension of the Irish Sea Basin into the Cheshire Basin and Worcester Graben. For this reason, a brief account of the onshore basins is included here.

The Cheshire Basin (Figure 24) is the low-lying area of Permian and Triassic rocks, extending from the Pennine uplands in the east to the Welsh Border hills in the west. South of Kidderminster, the basin narrows to a tongue extending through Worcester towards Gloucester. This series of fault-bounded and fault-controlled basins is believed to form part of the rift valley systems which are superimposed upon the North-West European Continental Shelf (Figure 9).

Rocks of Upper Carboniferous age (including the Coal Measures) form a broken and discontinuous ring around the Cheshire Basin. Coal Measure rocks originally formed a continuous sheet across central England as far west as the Welsh mountains. It is probable that they still floor much of the Cheshire Basin although ribs of older rocks may, in places, protrude up into the Permo-Triassic strata. The Hercynian earth movements which occurred at the end of the Carboniferous period, uplifted and folded the strata and resulted in an extension of the Welsh Mountain massif and produced the upland spine of the Pennines.

The Permian and Triassic rocks were primarily deposited in an arid, terrestrial environment, rather than in marine basins.

Faults define the present Permo-Triassic basins and probably controlled the development of the basins during accumulation of the sediments. The products of erosion from the Hercynian uplands collected within these fault-bounded intermontane basins. The sediments rest unconformably on the underlying surface of erosion.

Permian rocks are recognised with certainty only in the Manchester area, where a thick basal sequence of sandstones is overlain by a tongue of marine sediments, *the Manchester Marls*, containing a Permian fauna. It may be presumed that this temporary marine invasion proceeded from the north along the eastern margin of the basin, although its precise extent beneath the basin is unknown. Elsewhere in the basin, the rocks form an unbroken sequence of dominantly red non-marine rocks, some of which may be equivalent in age to the Manchester Marls. However, it is impossible to draw a precise boundary between Permian and Triassic strata and the whole sequence has long been known as the New Red Sandstone. Nevertheless, it is probable that rocks deposited during the Permian period do, in fact, extend across much of the Cheshire Basin.

A simplified Permo-Triassic succession in this area can be regarded as follows.

Triassic

	Rhaetic	
	Keuper	red mudstones (marls) with salt beds and sandstones in the lower part
	Bunter	sandstones and pebble beds.
Permian		
	Manchester Marls	
	Collyhurst Sandstone: and basal breccias.	

It is not surprising that during erosion of the landscape of central England following upon the Hercynian earth movements, that breccias and pebble beds were laid down in parts of the central England basins. Indeed, the faults which delimited those basins probably continued to move during deposition and thus breccia fans were concentrated along the fault scarps. At the northern margin of the Cheshire Basin, the Collyhurst Sandstone contains breccias and pebbles at the base and the unit thickens and thins across fault lines, suggesting contemporaneous control of deposition. The sandstones, which may be up to 800 ft thick, are overlain by the Manchester Marls, a sequence of marls with thin limestones, both of which contain characteristic Permian marine fauna.

The Triassic rocks were deposited over a wider area of the Cheshire Basin and probably extended further southwards along the narrow Worcester Graben. They were also laid down in more regular and uniform layers than the underlying Permian strata.

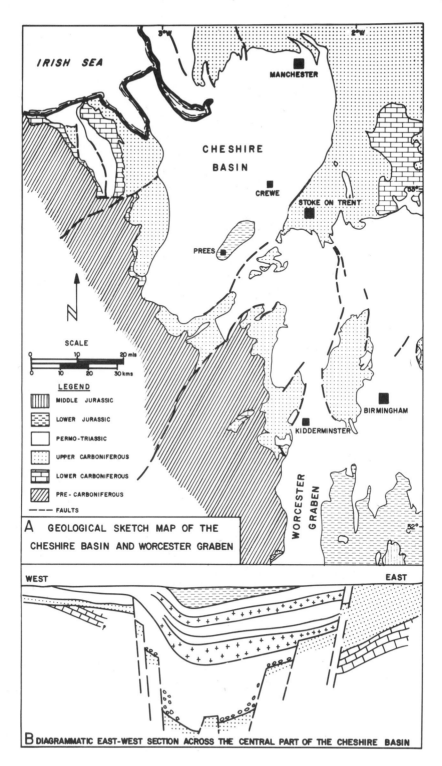

Figure 24
Geological sketch map and cross-section of the Cheshire Basin.

The topographic relief had been gradually subdued by erosion so that the Hercynian uplands had been reduced to a low peneplane in Triassic times.

The thick Bunter (approximately Lower Triassic) sequence comprises predominantly red sandstones, many of which show the rounded, polished sand grains and other features typical of desert dune deposits. However, the Bunter pebble beds and many of the upper sandstones also exhibit the characteristics of water deposition. The pebbles were deposited as fans and river valley deposits resulting from northward flowing rivers originating in the Welsh Mountains. Some of the more unusual pebbles suggest transport along a river system flowing northwards from Cornwall or Brittany. Certainly the pebbles decrease in number and abundance northwards across Cheshire and Lancashire.

The Keuper (Upper Triassic) deposits extend beyond the limits of Bunter deposition and, in places, extend onto the strongly folded pre-Permian rocks. In general, the lower part of the Keuper sequence is comprised of grey and mottled red sandstones. Suncracks, reptile footprints and other features all suggest a continuation of arid conditions with intermittent rivers and shallow lakes.

The Keuper Marl is a uniform and widespread unit which attains a thickness of several thousands of feet in the centre of the Cheshire Basin. The unit is composed for the most part, of red and chocolate coloured mudstones. Beds of evaporite are present in parts of the Keuper Marl sequence. In particular, there are two thick developments of salt beds in the central part of the Cheshire Basin. The Geological Survey Wilkesley borehole (see Figure 24) penetrated a combined thickness of 2,000 ft of salt beds. These, together with associated gypsum beds, were probably deposited during rapid and shallow marine invasions onto the low-lying Keuper plain. Subsidence of the land together with repeated phases of evaporation allowed thick salt beds to accumulate. The last event in Triassic times, was a widespread invasion of the sea and the deposition of the thin Rhaetic marine mudstones and limestones.

Various estimates have been made of the total thickness of post-Permian strata in the deeper parts of the Cheshire Basin. In the Institute of Geological Sciences Regional Handbook, an estimated thickness of 8,600 ft of New Red Sandstone is given for the axial portion of the basin. Certainly this appears to have been substantiated or exceeded by an oil boring at Prees (see Figure 24) during 1972-73. This well is believed to have drilled to a depth of 12,500 ft. Allowing for a thickness of Jurassic strata at the top of the hole and penetration into pre-Permian beds at the bottom, it would not be unreasonable to assume a thickness of 10,000 ft for the New Red Sandstone. An earlier deep boring near Formby, penetrated more than 5,500 ft of Permian and Triassic red beds and then entered the Upper Carboniferous (Millstone Grit). The well was stopped when it entered Lower Carboniferous Limestones at a depth of 7,500 ft.

The search for oil and gas within the Cheshire Basin results from geological conditions similar to those in the southern North Sea. In general, the Permo-Triassic red bed sequences contain few organic-rich source beds. This is also true in the southern North Sea, where it is generally accepted that the large natural gas accumulations discovered in recent years in Permian desert sandstones, have their source in the older Carboniferous Coal Measures. A similar situation could exist in the Cheshire Basin. However, in the North Sea, a seal for the Permian gas is provided by the Zechstein (Upper Permian) evaporites immediately overlying the sandstones. No such thickness of Permian evaporites occurs in Cheshire, where in the search for hydrocarbons, much could hinge on the thickness, nature and extent of the Manchester Marl in the subsurface and, in particular, on its ability to seal any potential trapping situation.

Seepages of oil are common in the coalfields surrounding the Cheshire Basin and are found at the surface in several places. For example, Coal Measures sandstones impregnated with tarry oil occur in Coalbrookdale. Of particular interest are the seepages at Formby, near the West Lancashire coast. Here oil was produced for some years from shallow wells drilled in oil impregnated Keuper Sandstones at the surface. Unfortunately, the deep source of this live flow of oil has so far been sought without success.

Selected Reading

BOTT, M P H and YOUNG, D G G	1971	Gravity measurements in the north Irish Sea. Q Jnl geol Soc Lond Vol 126.
BACON, M and McQUILLIN, R	1972	Refraction seismic surveys in the north Irish Sea. Q Jnl geol Soc Lond Vol 128.
DOBSON, M R et al	1973	**The geology of the southern Irish Sea. Rep Inst geol Sci 73/11.**
Institute of Geological Sciences	1969	Central England HMSO
Institute of Geological Sciences	1971	Northern England HMSO
PATTISON, J SMITH, D B and WARRINGTON, G	1972	A review of late Permian and Early Triassic Biostratigraphy in the British Isles. The Permian and Triassic Systems and their mutual boundary. Ohio University.

9 Northern Ireland, South West Scotland

9.1 Introduction

The Midland Valley of Scotland is an essentially downfaulted trough bounded to the south by the Southern Uplands Fault and to the north by the Highland Boundary Fault. The trough extends extends from the North Sea across Scotland through the Firth of Clyde and into Northern Ireland (see Figure 25). Formed initially as a Late Devonian structure it controlled much of the subsequent Late Palaeozoic and Mesozoic sedimentation particularly at the southern end in the Firth of Clyde and Northern Ireland.

Although the surface extent of Mesozoic and younger sediments in Northern Ireland takes the form of a broad belt stretching from Strangford Loch northwestwards towards Loch Foyle, the major accumulations of sediment are restricted to a pair of deep, fault-bounded Caledonian trending troughs, the *Magee* and *Portmore Basins,* separated by an intervening ridge of metamorphosed Precambrian rocks. This ridge, along with much of the flanking basins is largely obscured beneath a thick cover of Tertiary plateau lavas and forms a structurally consistent southwestern extension of the Precambrian basement exposed on the Mull of Kintyre. To the southeast the Magee Basin follows the line of the Midland Valley Trough and extends offshore beneath the Channel and Firth of Clyde as a submerged basin. North of the basement ridge, the Portmore Basin extends northeastwards from Loch Foyle towards the Sound of Jura forming a similar submerged deep sedimentary trough between the island of Islay and the Kintyre peninsula.

Although the principal development of the Midland Valley as a fault-bounded depression took place in Mid- to Late Devonian times, minor movements, associated with considerable volcanic activity, continued to take place throughout the Carboniferous. At the same time a shallow shelf sea invaded the depression, depositing marine and near-shore lagoonal sediments over both central Scotland and Northern Ireland. In Scotland, the Carboniferous was dominated by a continuous influx of erosional debris from the adjacent land areas of the Highlands and Southern Uplands, and in consequence is characterised by a sequence of sandstones, shales, thin limestones and coal seams. It is of note that some of the shale bands are rich in hydrocarbons and gave rise to a profitable oil shale industry during the early part of this century. Southwestwards over Northern Ireland in the Carboniferous, a clear shallow shelf sea dominated the Magee Trough

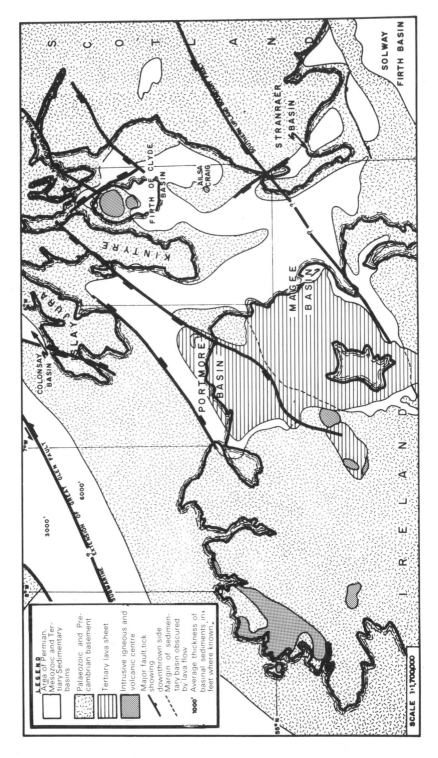

Figure 25
Geology Map of Northern Ireland and the Firth of Clyde.

depositing a thick limestone sequence similar to that occurring in central and southern Ireland. North of the Highland Boundary Fault extension however, coarse land-derived material again influenced the shallow marine sediments of the Londonderry-Donegal region.

Following a period of continental conditions the sea once again invaded the region for a time during Upper Permian — Lower Triassic times, and extended westwards from Rockall Trough to flood parts of the Firth of Clyde, Northern Ireland and the north Irish Sea. Elsewhere in the Firth of Clyde and over parts of Northern Ireland continental conditions still persisted with the deposition of thick sequences of desert sandstones and conglomerates. A return to widespread marine conditions took place in the Jurassic when the Liassic Sea extended westwards across the British Isles to link up with the Rockall Trough seaway. Fragmentary evidence suggests that much of the lowland regions of Northern Ireland, the Firth of Clyde and the Sea of Hebrides were originally overlain by a thin horizon of Jurassic marine sediments, much of which became rapidly worn away during the subsequent period of Lower Cretaceous uplift (Late Kimmerian earth movement phase). Lower Jurassic sediments are known from borehole evidence to exist as a thin layer of shales and limestones throughout much of Northern Ireland. However, in the Firth of Clyde the only evidence for the northward extension of this Lower Jurassic horizon is represented by small fragments preserved within the volcanic complex of Arran. Elsewhere over the floor of the Firth, south of Arran, there is as yet no evidence of any Jurassic or Cretaceous strata.

By the beginning of the Upper Cretaceous, a shallow embayment of the Chalk Sea extended eastwards across the shelf from Rockall Trough submerging much of the Inner Hebrides, Firth of Clyde and Northern Ireland region. A thin layer of basal sandstone overlain by chalk was laid down unconformably across the denuded surface of Jurassic, Triassic or older rocks. As with the Jurassic sediments, much of this chalk layer has subsequently been removed by erosion, and indications of its former extent are now only provided by a few isolated remnants throughout the region, such as the tiny blocks preserved within the centre of an extinct volcano on the Isle of Arran. However, in Northern Ireland the Upper Cretaceous is still present in zoth the Portmore and Magee basins as a layer of hard white chalk, overlying a basal unit of greensand. Borehole data indicates that in places this succession may reach thicknesses of up to 500 feet.

During the outburst of volcanic activity which accompanied the initial rupture of Greenland from Rockall, large volcanic vents broke through to the surface in both Northern Ireland and the Firth of Clyde, and gave rise to outpourings of extensive plateau-like sheets of basaltic lava.

9.2   Geophysical and Borehole Data

Much of the onshore extensions of the Permian — Mesozoic basins in Northern Ireland are obscured beneath a thick cover of Tertiary lavas. However, it has been possible to compile a fairly detailed knowledge of both the age and thickness of their major sedimentary infill from a number of deep boreholes drilled in the region, in conjunction with limited data from the narrow strip of sediments exposed beneath and around the margin of the plateau lavas. With the exception of a relatively limited magnetic survey north of Donegal, no geophysical data is available for the marine area between Northern Ireland and Scotland. In contrast however, extensive geophysical work in the form of gravity, magnetic and seismic surveys has been carried out over the Firth of Clyde, and an interpretation of these, backed up by a number of deep-water drill holes, shows that the southern part of the Firth is underlain by two thick Permo-Triassic sedimentary basins. Similarly, to the northeast of Arran, (although there is a lack of good borehold data here), it seems likely that Permo-Triassic sandstone may also provide the major infill to a smaller fault-bounded basin.

Both gravity and magnetic data across the Firth show a marked belt of high values believed to represent an upfaulted ridge of basement. This extends southwards from Kildonan Point on Arran, through the granite island of Ailsa Craig towards the Lower Palaeozoic schists of the Southern Uplands, and is flanked on both sides by a broad zone of low gravity values, which extensive sea-floor drilling has confirmed as indicating a pair of deep basins containing sediments of Permo-Triassic age. The Permo-Triassic basin fill rests unconformably on Devonian sandstones, or possibly locally on a Carboniferous basement.

In general, the Permo- Triassic sediments cored in the western basin are lithologically similar to those exposed on southern Arran, and show a succession of fine-grained sandstones or marls with thin bands of evaporites. However, drilling in the vicinity of the eastern low indicates that the Permo-Triassic infilling this basin is chiefly composed of wind-blown and therefore continentally-deposited sands. From the percentage of marine marls and evaporites present in the boreholes it appears that, while shallow marine and near-shore conditions persisted over the southern part of the Firth during the Permo-Triassic, the north and eastern regions were still experiencing continental conditions.

Geophysical data suggests that the eastern basin formed a small confined area of continental sedimentation while the western basin represents part of a much larger structure which extends southwards beyond the Firth of Clyde to both link up southwestwards with the broad Magee Basin of Northern Ireland, and southeastwards with the narrow fault-bounded trough of the *Stranraer Basin*. Geophysical data suggests that the Stranraer Basin may contain over 14,000 ft of Permo-Triassic sediments.

9.3   Stratigraphy

The Midland Valley and the Loch Foyle troughs were already in existence as major depressions by the end of the Devonian period. It seems probable, therefore, that much of the basement underlying the Magee — Firth of Clyde Basin and the Portmore Basin will be formed by Carboniferous rocks (unless removed by subsequent Late Palaeozoic erosion). In regions where the Carboniferous is absent, such as in parts of the Firth of Clyde, the underlying basement is likely to consist of either Devonian sandstones or older metamorphosed Lower Palaeozoic and Precambrian rocks.

In Northern Ireland, Permo-Triassic sediments are fairly widely distributed both in the Magee-Portmore region and are also represented to the south in a small isolated, fault-controlled basin at Kingscourt, County Cavan. Although the Upper Permian sediments here are notably more marine in nature, by comparison with the continentally deposited sandstones and pebble beds exposed on Arran to the north, it appears that by Triassic times a much more uniform environment of deposition existed over the whole region.

The Permian of Northern Ireland shows a characteristic three-fold stratigraphic division which comprises a basal unit of coarse sandstones, pebbles and rock fragments, overlain by an intermediate unit of marine limestones and then by an upper unit of mudstone. This upper mudstone layer contains bands and lenses of evaporite deposits up to 15 feet in thickness, and as such is somewhat similar to the rocks encountered in shallow drillholes over the southern half of the Firth of Clyde. The overlying Triassic forms a more uniform continental sequence characterised by sandstones, sandy mudstones, and occasional bands of clayey mudstones. The entire succession within the centre of the basin has an average thickness of 2,000 feet and the occurrence of fossilised sedimentary features such as mudcracks, ripple marks and rare reptilian footprints suggests that they were laid down in a fairly shallow-water or shoreline environment.

With the exception of minor Jurassic and Cretaceous fragments preserved on Arran, the youngest identifiable sediments within the Firth of Clyde are of Triassic age. As a result, the entire stratigraphical picture of the Jurassic and Cretaceous sediments, which infill parts of the Magee-Firth of Clyde Basin and the Portmore Basin, has been built up from observations of the relatively restricted outcrop sequence in Northern Ireland.

During the Middle and Upper Jurassic much of Northern Ireland and southwestern Scotland was emergent, while the lower areas were submerged beneath a shallow-water sea. With the retreat of the sea, and the gradual emergence of these lower areas in Upper Jurassic and Early Cretaceous times, much of the earlier deposited sediments were removed by erosion. As a result, the entire Jurassic sequence is now only represented by a thin (roughly 500 ft thick) marine unit of shales and limestones. Similarly, most of the Lower Cretaceous is missing and the Upper Cretaceous sequence is composed of hard white chalk underlain by Greensand which rests directly on an eroded surface of Lower Jurassic, Permo-Triassic or even older basement rocks. Borehole evidence suggests that the maximum thickness of the chalk layer deposited across Northern Ireland was in the order of 450 — 500 feet, and although now considerably eroded, it is believed to have originally been a widespread sheet which extended across much of the region.

Although Tertiary sediments are generally only poorly developed over Northern Ireland, and appear to be entirely absent across the shelf between Ireland and the Firth of Clyde, they do reach a local thickness of over 1,000 feet in the Lough Neagh Basin. Here a sequence of clays and sands believed to be of Miocene or Pliocene age are resting directly on the surface of the plateau basalts.

Selected Reading

CHESHER, J A et al   1972   I G S marine drilling, with m.v. Whitehorn in Scottish waters 1970 — 1971.
Rep Inst Geol Sci 72/10.

McLEAN, A C et al   1970   Gravity, magnetic and sparker surveys in the Firth of Clyde.
Proc Geol Soc Lond Vol 1662.

10 West Scottish Basins

10.1 Introduction

A chain of deep asymmetrical basins extends northwards across the submerged continental shelf west of Britain from the Irish Sea through the Sea of Hebrides and the Minches, and northwards across the shelf lying to the west of the Orkney and Shetland Islands (see Figure 9). The basins follow a general northeastward trend, and are believed to represent tensional features initiated during the Early Mesozoic in response to an abortive attempted separation of the Greenland — Rockall Plate from the West European Plate along the line of Rockall Trough and Porcupine Seabight. Following the development and reactivation of a linear series of fractures during the early stages of plate separation, certain sections of the crust adjacent to the fracture lines began to subside. These subsided sections formed wedge-shaped troughs which subsequently became infilled with up to 15,000 ft of Mesozoic and younger sediments.

Geological and geophysical evidence in this area shows that the structure of the shelf is controlled by four principal fault systems — the Minch Fault, the Camasunary-Skerryvore Fault and the Great Glen Fault to the west of Scotland, and a continuation of the Minch Fault and West Hebridean Fault to the north (see Figure 12). These faults form the margins to a number of deep sediment-filled troughs floored by downthrown ancient basement rocks of Precambrian and Lower Palaeozoic age, thus their history of development is of considerable importance to the understanding of the type and thickness of the sedimentary section preserved within the basins.

Recent work over the Sea of Hebrides now shows the Minch Fault to be a complex system of interacting faults which downthrow basement towards the southeast as opposed to the earlier belief that it behaved as a single transcurrent fault. Considerable transcurrent movement did in fact occur during the early history of the tensional stresses associated with plate separation. Subsequently this line of weakness became reactivated throughout the Mesozoic and Tertiary, when subjected to increasing tensional stresses, which gave rise to repeated vertical movements leading to the gradual increase in depth of the adjoining basins in the Minch.

Similarly the line of the Camasunary — Skerryvore Fault to the east, of comparable trend and vertical displacement to the Minch

Fault, appears to have undergone a major phase of vertical movement during the Jurassic, which was followed by repeated movements of lesser magnitude throughout the Tertiary. Evidence of any earlier movements is obscure, although it is possible that following the pattern of the adjacent Great Glen and Minch Faults, this fault may also have been initiated as a transcurrent fracture in Upper Palaeozoic time.

The line of the Great Glen Valley in Scotland follows a major tectonic fracture along which the north of Scotland has slid and is still sliding against the south. At present the direction of this relative movement is in dispute. Offshore the fault extends northwards closely following the east coast of Scotland and the Orkneys to intersect the Shetland Isles as the Walls Boundary Fault. Southwards the fault divides into two; one branch passes to the north of the Isle of Colonsay bounding a deep Mesozoic and Tertiary filled basin, and the other branch passes through Islay to eventually link up with the Lennan Valley Fault in Ireland (see Figure 12). Until recently, geological evidence had been interpreted as indicating that in pre-Upper Carboniferous times the northern part of Scotland moved northward by some 65 miles with respect to southern Scotland. However, with considerably more data available, it now seems likely that motion in the opposite sense may have been dominant with repeated movements occurring along the fault line from Upper Palaeozoic times onwards.

As a result of the periods of vertical displacement along these faults, the West Scottish Basins began initially to develop as fracture-controlled sag structures during the Permo-Triassic. Continued crustal stresses accompanying the extension of the mid-Atlantic spreading system northwards towards Iceland resulted in the basins undergoing repeated phases of downtilting throughout Upper Mesozoic and Tertiary times.

In Late Carboniferous — Early Permian times a period of widespread uplift affected much of Western Europe including parts of Scotland and England, and led to a general increase in aridity and the development of a *desert type* environment of deposition. During the uppermost Palaeozoic to Early Mesozoic time interval, the downfaulted basins became the sites of fluviatile and lacustrine deposition with accumulation of silts, sands and gravels washed down from the surrounding higher ground building up across the trough floor. Later, during the Mesozoic, these continental deposits were submerged beneath the advancing Lower Jurassic sea, and throughout the subsequent Jurassic, Cretaceous and Tertiary period, marine conditions prevailed to the north and west of the Scottish mainland.

In the Minch — Sea of Hebrides region, four deep, sediment-filled basins have been downfaulted into the continental shelf. Two of

Figure 26
Geology Map of Western Scotland and the Hebrides

these, the *North Minch Basin* and the *Little Minch — Sea of Hebrides Basin* form a pair of elongate troughs closely paralleling the eastern coast of the Outer Hebridean Islands, and separated from each other by NW-SE trending ridge of basements rocks (Figure 26). The third and fourth, known respectively as the *Inner Hebrides Basin* and *Colonsay Basin* form a second paid of sub-parallel troughs to the southeast, in the vicinity of Mull. Mull.

To the north of Scotland, the same NE-SW trend is displayed by the *West Shetland* and *Sule Sgeir Basins* lying west of the Orkney and Shetland Islands (see Figure 28 A). The first forms a deep asymmetrical trough, sharply downfaulted to the east against the basement rocks of the Shetland Platform, while the second may be controlled by the northeastward continuation of the Minch Fault along its western margin.

10.2   **Geophysical and Sample Data**

The earliest indications of thick sedimentary rocks being present in the continental shelf region to the north and west of Scotland came from the analysis of gravity measurements across the area. These data showed that the shallow marine area of the shelf was dominated by a pattern of sharply defined, elongate, low gravity areas. This in itself suggested that parts of the shelf were not underlain by the ancient Precambrian basement or thick Devonian sandstones typical of the Orkney, Shetland and Hebridean Islands and the north and northwest Scottish mainland coast, but were probably formed by considerably younger sediments. By the late 1950's these low gravity areas were correctly interpreted as being due to a series of partly fault-bounded sedimentary basins, and the then rather broadly defined position of the basins served as a basis for all the subsequent more detailed geophysical investigations. Seismic reflection data in conjunction with sea-floor sample data, have been able to define the limits of the basins more accurately, and also provide some evidence for the age, thickness and nature of the infilling sediments. In recent years many thousands of miles of high quality seismic reflection lines have been acquired by the oil industry in this region.

West of Scotland a steep gravity gradient appears to closely follow the line of the Minch Fault, separating the high gravity basement rocks of the Outer Hebridean Islands from a region of marked gravity lows centred over Northern Skye and the Little Minch, and to the east and northeast of Stornoway, Lewis. To the southeast, two further lines of marked gravity lows are centred in the offshore region between Mull and the high gravity Coll — Tiree basement ridge, and in the vicinity of Colonsay respectively.

Also to the north of Scotland, a strong NE-SW trending belt of low gravity, some 30 km broad and 100 km long, was detected lying to the west of the Shetland Islands. This is now termed the West Shetland Basin. To the west of the Orkney Islands a roughly circular zone of low gravity similarly defines the Sule Sgeir Basin.

The most revealing geophysical work which has led to a better understanding of these basins comprises a fairly closely spaced network of seismic reflection profiles shot across the marine area of the Inner Hebrides and a few well placed long reflection profiles across the shelf to the north of Scotland. Subsequently, further seismic programmes have been run commercially across the West Shetland and Sule Sgeir Basins by individual seismic and oil companies who are interested in the hydrocarbon prospects of the area. Although such data will have led to a fairly complete understanding of the geology and structure of the basins, the results of their findings are naturally still confidential.

Geophysical data show that the Little Minch — Sea of Hebrides Trough reaches a maximum depth in the vicinity of northern Skye, where the thickness of the infilling sediment is in the order of 7,500 — 9,000 ft. Seismic profiles across the trough show it to be not only strongly asymmetrical in cross-section (with the maximum sag occurring along the western margin adjacent to the Minch Fault) but also to exhibit an overall downward tilt towards the northeast so that the southwestern end of the basin shallows and eventually wedges out against the Minch Fault. The main infill of the trough has been identified from the appearance of regular bedding and velocity values observed, to be of probable Mesozoic age. The infill is subdivided into a thin upper unit of lower velocity sediments and a thicker lower unit of somewhat higher velocity sediments overlying metamorphic basement. The upper unit is found to reach a maximum thickness of approximately 1,000 ft northwest of the Isle of Rhumm, thinning laterally towards the basin margins, while the lower unit reaches much greater thicknesses, in the order of 8,000 ft, in the vicinity of northwest Skye, again thinning towards the basin margins.
The average interval velocity of the sediments in the upper unit is 8,200 ft/sec and in the lower unit is 11,500 ft/sec.

The Inner Hebrides Basin is narrower than the Little Minch — Sea of Hebrides Basin but gravity and seismic reflection data show the maximum thickness of infilling sediments to be of a similar order. Likewise, the deepest part of the basin is developed along its western margin adjacent to the Camasunary — Skerryvore Fault line. Eastwards the sediments thin to overlap across the Palaeozoic basement rocks of Mull and the mainland coast (Figure 27).

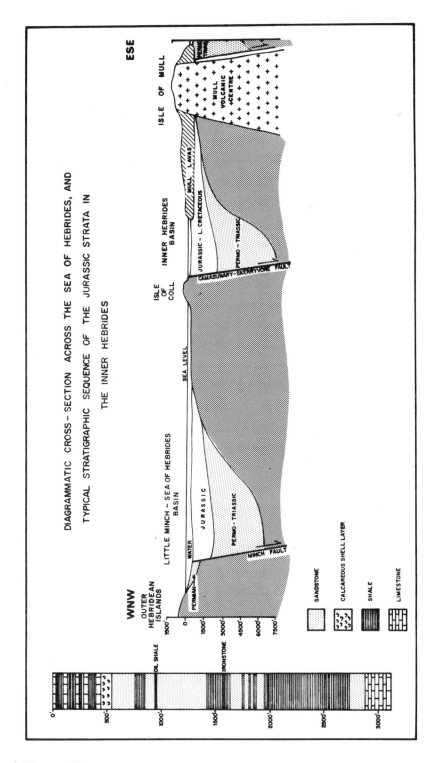

Figure 27

Diagrammatic cross-section across the Sea of Hebrides and typical stratigraphic sequence of the Jurassic strata in the Inner Hebrides

To the southeast the Great Glen Fault can be identified as forming the north-western margin of a further sedimentary trough containing some 6,500 ft of young sediments: gravity data over the southern part of the Sea of Hebrides suggest that this basin may extend southwestward following the line of the Great Glen Fault to eventually link up with the sediment-covered margin of Rockall Trough.

To the north of Scotland, seismic evidence across the West Shetland Basin shows the infilling sediments to comprise a thick lower succession of southeastward dipping sediments which reach a maximum thickness along the fault-bounded eastern margin, and an overlying thin succession of near-horizontal or gently westward-dipping sediments.

To the east of the Orkney Islands seismic reflection data show the Sule Sgeir basin to be asymmetrical, bordered to the northwest by the major downthrow of the Minch Fault and infilled by a thick sequence of northwesterly dipping sediments, covered by a thin layer of near-horizontal younger sediments. The southeasterly limit of the basin is also controlled by faults which bring the young sediments against the Precambrian and Palaeozoic basement of the Sule Sgeir ridge and the Shetland Platform.

To the west and parallel to the Outer Hebridean Islands, gravity data indicate the presence of a narrow fault controlled basin, the Outer Hebrides Trough. The data suggest that this basin is deeper in the north adjacent to the Isles of Lewis and Harris, and shallows southwards in the vicinity of Barra and South Uist. Throughout the length of this trough the eastern margin is marked by a rapid change in gravity level by comparison with that observed over the western margin, which suggests that the trough also has an asymmetrical form but in an opposite sense to that of the basins in the Minches and Sea of Hebrides.

10.3   Stratigraphy

West of Scotland, the Outer Hebridean Islands of Lewis, Harris, Uist and Benbecula are composed of ancient, highly metamorphosed Precambrian rocks similar to those forming the adjacent mainland of Scotland and parts of the Shetland Isles. However, throughout the Inner Hebridean Islands of Mull, Eigg, Skye and Raasay and along the west coast of the mainland in areas such as Ardnamurchan, Morvern, Applecross and Gruinard Bay, small patches of Mesozoic rocks have been preserved either by downfaulting or as a protected sequence beneath the thick overlying layers of Tertiary lavas. These fragments represent the exposed margins of the offshore downfaulted basins, and as such provide valuable geological control for the age and lithology of the infilling sedimentary sequence.

Permian and Triassic sediments containing potentially excellent reservoir horizons are exposed as a sequence of coarse continentally-derived rocks comprising siltstones, sandstones and pebble beds. The sequence rests directly on the irregular basement floor of the troughs, and where exposed can be seen to locally reach considerable thicknesses, eg 1000 ft on the east shore of the North Minch at Gruinard Bay, 10,000 ft at Stornoway on Lewis and up to 6,600 ft in South Mull. Marine limestones and sandstones of uppermost Triassic and basal Jurassic age are known only in the southern part of the area, in western Mull, where they are rapidly overstepped by younger Jurassic marine sediments. However, concurrent with the deposition of these marine strata, further north sediments were still being laid down in a continental environment.

Jurassic sediments attain a thickness of 3,000 ft in the region of northern Syke, Raasay and Applecross where they are characterised by a thick series of sandstones, interbedded with hard, calcareous clays, and black oil-shales. The Middle Jurassic is locally dominated by a deltaic succession of shales, thin limestones and sandstones. Elsewhere, an incomplete Jurassic sequence is present on Mull and as minute exposures on the Isle of Arran, the Shiant Isles and on the mainland at Gruinard Bay.

A major break in sedimentary deposition occurred during the Late Jurassic and Early Cretaceous period throughout the Hebridean Province. By Upper Cretaceous times a thin (up to 70 ft) but widespread sandstone and chalk sequence was deposited unconformably across the eroded surface of older Jurassic, Triassic and Permian rocks. Erosion has removed much of this Upper Cretaceous deposit, but a few scattered remnants have been preserved on Mull, Arran and around Loch Aline. These are composed of well-rounded, highly porous sandstones capped by a thin succession of chalk. Tertiary sediments appear to be absent throughout the Hebridean region, although they are believed to be fairly extensively developed to the west of the Shetland Platform. However, extensive areas of the Inner Hebridean Islands and the floor of the Minches and Sea of Hebrides are covered by Tertiary igneous rocks. Marine sampling across the basins west of Scotland have yielded sea-floor cores of Triassic, Jurassic and Cretaceous age.

The Little Minch — Sea of Hebrides Basin forms an extensive trough extending southwards from the Isle of Lewis (57°55′N) beneath the Tertiary lavas of Skye and the Little Minch into the Sea of Hebrides, where it narrows to finally pinch out some 18 miles south of Berneray, Outer Hebrides (see Figure 26). To the northwest the basin margin closely follows the coastline of the

Outer Hebrides, although the main thickness of infilling sediments lies immediately east of the Minch Fault. Southeastwards the basin floor shelves gently onto the Precambrian and Lower Palaeozoic ridge of Rhum, Coll and Tiree. At the northern end of the basin, the sediments abut sharply against the NW-SE trending basement ridge which acts as a barrier between this basin and the North Minch Basin to the north.

Rock samples taken from the sea-floor show that Jurassic or Cretaceous rocks occupy the deep central part of the basin thinning towards the basin edges where the underlying older Permo-Triassic sandstones, silts and pebble beds are exposed. The thicknesses of Jurassic sediments on land suggest that rocks of this age are not likely to form the principal component of the basin infill. Correlation with seismic reflection data suggests that they only represent the relatively thin upper seismic unit recognised on the seismic profiles, while the underlying denser seismic unit is thought to represent a thick Permo-Triassic sequence (Figure 27).

To the east of the Isle of Lewis, the North Minch Basin follows the same lineament as the Little Minch — Sea of Hebrides trough, being similarly controlled by the line of the Minch Fault. Although to the north of the Outer Hebrides, the basin changes its trend to form a much broader basin which straddles the Minch Fault. At present, in comparison with the Minches and Sea of Hebrides to the south, geophysical work has been rather limited over this northern area. However, all the available evidence points to a similar pattern of sedimentation in both areas, and it is anticipated that the major infill to the North Minch Basin will comprise an 8,000—10,000ft thick section of continentally derived Permo-Triassic and marine Jurassic sediments.

To the southeast of the Little Minch — Sea of Hebrides Basin, and and separated from it by the NE-SW trending basement ridge of Precambrian and Palaeozoic rocks exposed on the islands of Rhum, Coll and Tiree, is the similar trending Inner Hebrides Basin (see Figures 26 and 27). This basin, controlled by the Camasunary Fault, extends southwards from southern Skye beneath and along the west coast of Mull towards the submerged Blackstones Tertiary igneous centre at 56°5′N. South of Blackstones the basin forms the northeastern part of a 2,000—3,000ft deep trough which extends across the continental shelf into the margin of Rockall Trough.

The asymmetrical nature of the Inner Hebrides Trough is not only evident from geophysical data but is also apparent from the sediments exposed onshore in southern Skye, some 8 miles to the east of the Camasunary Fault. Hereabouts, westward thickening Triassic and Lower Jurassic sediments rest directly on the eroded basement surface of Precambrian and Lower Palaeozoic rock. Marine samples from the seafloor in the northern part of the trough have yielded Jurassic sediments, but to the south in the vicinity of Mull, where much of the thickest part of the trough is overlaid by Tertiary lavas, it is thought that the sediments will be similar to the Triassic sandstone and pebble beds exposed on the Isle of Inch, off western Mull, and to the Lower Jurassic calcareous sandstones, limestones and shales exposed in southern Mull. Seismic reflection data across the basin show an infilling succession of similar seismic character to that identified in the Little Minch—Sea of Hebrides Basin, and is now believed to comprise a thick (7,000—8,000ft) lower unit of Permian and Triassic continental sediments overlain by a much thinner upper unit of marine Jurassic rocks.

Eastwards the Mesozoic sedimentary succession of the Inner Hebrides Basin thins considerably across Mull and appears to pinch out beneath the Tertiary lavas. Figure 26 shows that the line of the Great Glen Fault cuts across the southeastern coast of Mull and acts as the northern margin of a further deep asymmetrical trough to the southeast. This basin appears to be mainly infilled by Mesozoic rocks and overlaid by a southwestward thickening unit of Tertiary marine sediments (probably limestones and clays). A subsidiary finger of the basin extends southwards between the islands of Colonsay and Jura. Along its western margin it is strongly downfaulted against the metamorphic basement of Colonsay, while at its southern end it pinches out against the basement feature of Islay. Sea-floor samples over the basin show it to be infilled by sandstones and pebble beds of probably Permo-Triassic age.

The mainland coast of the northern part of Scotland, and the tiny offshore islands of Sule Skerry, Skerry Stack and North Rona, along with most of the Shetland Islands are composed of ancient Precambrian and Lower Palaeozoic basement rocks and as such provide no indication as to the nature of the sediments expected to infill the West Shetland and Sule Sgeir Basins. The only other rock type exposed in this area forms a belt of Upper Palaeozoic (Devonian) continentally deposited sandstones and pebble beds. These occupy a northeastward trending tract from the Moray Firth, through Caithness and the Orkney Islands, into the southern edge of the Shetland Islands. Although normally classified as basement because of their Palaeozoic age, these coarse sediments possess quite good reservoir properties, and under certain circumstances could act as reservoir beds for the accumulation of hydrocarbons. However, towards the west the succession thins and eventually pinches out against the basement surface of the Shetland Platform, and so is unlikely to provide any of the basal infill in the younger downfaulted West Shetland and Sule Sgeir Basins.

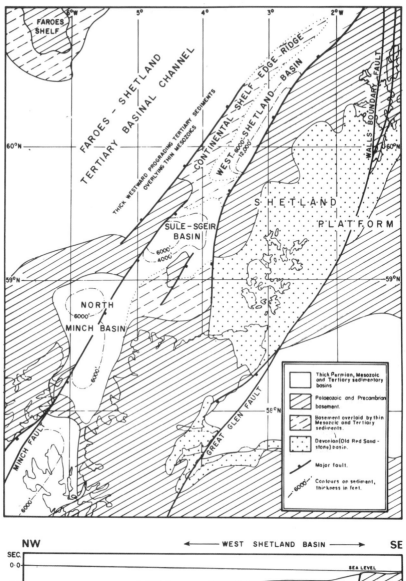

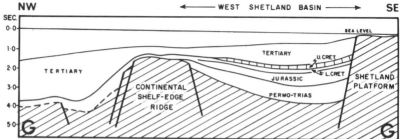

Figure 28

A   *Geological sketch map of the West Shetland area.*
B   *Section across the West Shetland Basin based on an existing seismic profile.*
For location see Figure 13.

The lack of publicly available borehole data over the West Shetland and Sule Sgeir Basins makes an interpretation of the infilling sedimentary succession still somewhat conjectural. In the West Shetland Basin the thick lower succession of south-easterly dipping sediments terminates abruptly against the fault-bounded eastern margin. Here a maximum of 20,000ft of sediments is preserved (see Figure 28B). Both the seismic appearance and character of this lower dipping sequence suggest that it is composed of Mesozoic continental and marine sediments of Permo-Triassic, Jurassic and Cretaceous age. The overlying thinner unit of younger sediments which rests unconformably on this dipping sequence is believed to be mainly of Tertiary age, although it may be partly composed of Cretaceous strata. If this latter suggestion were correct then a thicker (and more prospective) Jurassic section could be developed in the basin than is shown on Figure 28B.

Although the major sedimentary component infilling the Sule Sgeir Basin dips in the opposite direction to that within the West Shetland Basin, it is generally anticipated that the 10,000ft thick succession will be stratigraphically very similar in both basins. The overlying near-horizontally bedded sequence of probable Tertiary age thinly covers the northwestward dipping Mesozoic rocks to extend southwards and southwestwards beyond the limits of the basin as a shallow Tertiary trough linked into the North Minch Basin.

Selected Reading

BINNS, PE et al	1974	The geology of the Sea of Hebrides. Rep Inst geol Sci 73/74
CHESHER, JA et al	1972	IGS marine drilling with mv Whitehorn in Scottish waters 1970-71 Rep Inst geol Sci 72/10
FEIR, GD	1971	New seismic evidence of large scale faulting on the Shetland Hebridean continental margin. First European Earth and Planetary Phys Colloquim. Reading March 30 1971
GARSON, MS and PLANT, J	1972	Possible dextral movements on the Great Glen and Minch Faults in Scotland. Nature Phy Sci Vol 240 Nov 13.

11 Palaeogeography and Evolution of Western Britain

From Precambrian to Recent times the British Isles and surrounding marine areas have undergone a complex sequence of geological events during which a series of structural provinces and depositional basins have been created. As a result the geological history of the offshore region of western Britain and Ireland can only be fully understood when it is seen in relation to the broad pattern of the tectonic and palaeographic evolution of the entire Western Europe-North Atlantic province. The familiar shape and coastline of the British Isles has emerged only recently.

From the economic viewpoint, the rocks most likely to contain hydrocarbons are those deposited in Upper Palaeozoic, Mesozoic and Tertiary times. During this period such sediments were deposited in basins created and controlled by the following five major phases of earth movements.

1   Caledonain phase of mountain building. (Cambrian to Devonian).
2   Hercynian phase of mountain building. (Upper Devonian-Upper Carboniferous).
3   Kimmerian earth movements. (Lower Jurassic to Upper Jurassic).
4   Alpine phase of mountain building. (Tertiary, Oligocene to Pliocene).
5   Separation of the Greenland-American Plate from the Western Europe-Rockall Plate. (Lower Tertiary to present day).

The palaeogeographic conditions of the British Isles and surrounding continental shelf from Devonian times onwards, are considered in more detail than those of the earlier Precambrian and Lower Palaeozoic period.

Building of the initial structural framework in Western Europe began in Early Precambrian times (roughly 2,600 million years ago) with the deformation, folding and metamorphism of the ancient rocks which formed the stable platforms of the North-Western and Russian Baltic Shields (see Figure 10). Remnants of this early deformation can still be recognised in the Outer Hebridean Islands and parts of the northwestern coast of Scotland, and in scattered fragments throughout England and Wales.

During the long Precambrian and Early Palaeozoic eras, a broad trough existed between the already relatively rigid North-Western Platform and the Russian Platform providing a depositional centre into which were poured the erosional products of both flanking shields. The southern extremity of the trough stretched across Scotland, northern England, Wales and much of Ireland, and with the gradual approach of both shield areas during Lower Palaeozoic, Ordovician, Silurian and Devonian times, the powerful fold movements of the *Caledonain* phase, crumpled these basinal sediments and uplifted them into a chain of high mountains which stretched from Norway into southern Ireland and the Low Countries. The worn-down remnants of this ancient mountain chain still form important topographical features in these countries at the present day.

In Early Ordovician times the Scottish Highlands (including the Moine Thrust system of dislocation to the northwest) were slowly uplifted to form a major land region, while to the south the remainder of the British Isles was still submerged beneath a shelf sea. With the continued approach of both shield areas, similar palaeogeographical conditions persisted into the Silurian, although there was a general tendency towards shallowing of the shelf sea and the appearance of an emerging belt of islands through the Lake District and parts of the Irish Sea and St George's Channel. By Late Silurian and Early Devonian times, the strong Caledonian earth movements reached their climax creating one vast mountainous continent and forcing the shallow shelf sea to migrate southwards out of all but the southernmost portion of Britain.

With the withdrawal of the sea, continental conditions prevailed over most of the newly created land area. Rapid erosion of the high ground, along with further Late Caledonian earth movements, led to extensive downwarping of parts of the crust and the creation of a broad continental depression in the vicinity of southern Norway and the Orkneys and Shetlands, and a narrow downfaulted depression across the centre of Scotland (Midland Valley) linking the North Sea sedimentary region with that of the Sea of Hebrides and Northern Ireland. Within these newly formed basins, thick Devonian fluviatile and lacustrine sandstones were laid down. At this time the shoreline between the northern continent and the southward deepening sea stretched from southern Ireland across southern England into northern France, with typically marine limestones, shales and thin sandstones being deposited across the Celtic Sea, Cornubian Massif and Western Approaches region.

The emergence of the Late Devonian-Carboniferous (*Hercynian*) mountain chain through the Brittany-Normandy peninsula and central France changed the entire palaeogeographical pattern of southern Britain (see Figure 10). The mountains created a well-defined southern boundary to the existing Devonian Sea, and provided a rich source of continentally derived sandstones and mudstones. Initially the uplift of the Hercynian mountains pushed the edge of the sea northwards across the continent, and the Early Carboniferous was marked by the encroachment of the sea into continental depressions in northern England and Scotland, and the deposition of marine limestones over the more stable regions and marine shales in the more rapidly subsiding regions. During this period much of Wales and the southern Irish Sea formed a large island, (St George's Land) which persisted as an upland feature from Early Carboniferous times onwards. Similarly in the north, most of Scotland remained a major positive area and was flanked along its southern boundary by a broad coastal plain of lagoons, swamps and deltas in which beds of coal were laid down locally. How much of Ireland was submerged at that time is not known for certain, but the deep water sediments of the extreme southwest can be seen to pass northwards into shallow reefal limestones.

By Upper Carboniferous times deep water sedimentation was restricted to a narrow trough flanking the rising Hercynian chain. With the exception of the land areas formed by the Scottish Highlands and the Welsh-Midlands Island, interfingering shallow water and coastal swamp conditions prevailed over the majority of Britain and Ireland. These swamp conditions lead to the deposition of locally very thick coal-bearing sequenced throughout central and northern England, the northern Irish Sea, Midland Valley of Scotland, South Wales, southern England and the southern North Sea, and over wide areas of Ireland. Later crustal disturbances and erosion have meant that Coal Measures are only now preserved over parts of these areas. Nowhere is this more clearly demonstrated than in Ireland where Coal Measures are preserved only as small isolated localities.

Towards the end of the Carboniferous and beginning of the Permian, the swampy conditions of Britain largely gave way to a drier, arid or semi-arid, continental environment of low desert plains and newly created downfaulted depressions within which the products of erosion and weathering collected. To the south, the Hercynian mountains had been largely worn away and replaced by low hills in the Cornubian and Brittany area.

By Late Permian time, the sea once more invaded large regions of the Permian continent, flooding extensive depressional areas and creating two almost enclosed seas across the British Isles separated by a central barrier along the line of the Pennines (see

Figure 29). Throughout Late Permian time these seas remained shallow, varying considerably in extent and salinity, and this led to the deposition of sediments ranging from evaporites and salt to marls and limestones. To the east, the larger of the two seas, known as the *Zechstein Sea* covered much of the North Sea region and east cost of England extending eastwards into Holland, Denmark and northern Germany, and probably northwards via a narrow passage between Greenland and Norway into an open Arctic Ocean. To the west, the less extensive *Bakevellia Sea* occupied much of the northern Irish Sea region and the adjacent Lancashire, Cumberland and northern Irish coasts (see Figure 29); it extended northeastwards via a narrow passage between the northern Irish coast and Sea of Hebrides, into a broader basin occupying the present-day Rockall Trough region. With the lack of borehole data across Rockall Trough there is still considerable uncertainty about the existence of fairly open marine conditions here in Permo-Triassic times. However, it does seem likely that at that time a passage-way may have extended southwards from an Arctic sea, with one branch feeding the Zechstein Sea to the east, and another passing to the west down the line of Rockall Trough and possibly beyond between Spain and North America.

During the period of Late Permian sedimentation, the Bakevellia Sea appears to have periodically extended into the Cheshire trough system and Vale of Eden, and also to have been briefly connected to the Zechstein Sea to the east. However, truly marine conditions appear to have been fairly short-lived in the Bakevellia Sea, and as the sea shrank towards the end of the Permian, the shallow marine sequences of limestones, shales, evaporites and salt gave way to marginal marine and eventually continental sediments. By the beginning of the Triassic, arid and semi-arid continental conditions had again returned to much of west Britain and the former marine depressions became extensive centres for the deposition of coarse pebble beds and sandstones. With the continuous wearing away of the upland mountainous areas, the continental basins of west Britain and the Irish Sea became progressively infilled and eventually the Pennine barrier to the east was covered and the basins linked with the North Sea depositional centre.

Subsequently, minor earth movements in the Lower Triassic recreated this Pennine barrier and by the end of Lower Jurassic times the sea had again flooded the depressions formed in the Upper Permian, re-establishing shallow marine conditions for a short period over the north Irish Sea region before the return of more typical continental conditions. To the south, extensive fault-bounded depressions were becoming established in the Celtic Sea, Cardigan Bay and Western Approaches area forming the framework of the later Mesozoic and Tertiary troughs (see Figure 29). These depressions were the centre of thick sandstone

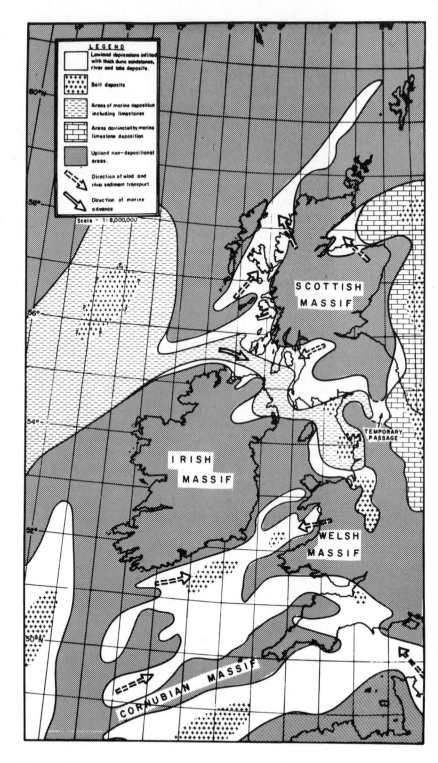

Figure 29
Palaeogeography of Western Britain during the Permo-Triassic.

deposition during the Permo-Triassic, and although not connected to the Bakevellia Sea which lay to the north of Wales, they were almost certainly briefly invaded by the sea from the east via the Bristol Channel depression during later Triassic times. This led to the extensive deposition of evaporites and salt beds offshore in the Celtic Sea and onshore in Somerset. Similar, although somewhat narrower, fault-bounded troughs existed through Worcestershire and Cheshire to the east of the St George's Land Massif, and to the north of the Bakevellia Sea between the Scottish Highlands and the Outer Hebridean Islands. Marine invasions of these troughs were generally rare, and to the north appear to have been entirely absent, with sedimentation being largely restricted to river-laid and lake deposits.

The end of the Triassic marked the end of the continental and shallow marine regime which had persisted over the British Isles since the Devonian. The uppermost Triassic and earliest Jurassic period was accompanied by a phase of minor earth movements (*Early Kimmerian Phase*) which caused the Tethys Sea of southern Europe to spread northwards, across the denuded surface of the Hercynian mountains into northern Europe and Britain. Associated downwarping and local crustal movements led to much of Britain being submerged beneath the advancing Liassic Sea while the higher upland areas of the Cornubian Peninsula, Wales and the Scottish-Pennine ridge remained emergent as extensive positive island features (see Figure 30). How much of Ireland was covered by this sea is difficult to determine. Although parts of northwestern Ireland were certainly flooded, recent work in the Irish Sea suggests that the *Welsh Massif* may have periodically stretched westwards to join a *Central Irish Massif*, thereby creating one large island which acted as a partial barrier between the new marine Liassic conditions of Northern Ireland and the Sea of Hebrides from those of the Celtic Sea and St George's Channel.

Lower Jurassic times also saw the start of separation of the West European Continental Plate from the North American Plate in the vicinity of the Spanish coast. Although it is unlikely that actual rupture of the two continents and the creation of new oceanic crust had occurred at such an early stage, much of the low relief regions between Rockall and the Danish-Belgium coast were almost certainly affected by the northward spread of a shallow shelf sea.

South of the Irish-Welsh Massif marine conditions spread into the Western Approaches and Channel depressions from two directions. One invasion came from the Tethys Ocean via northern France to the southeast; the other, from the newly established North Atlantic seaway to the southwest. North of the Cornubian Peninsula, conditions may have been a little more

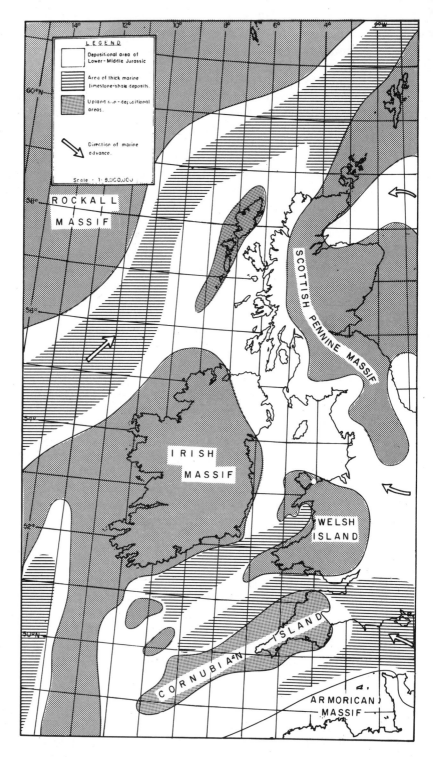

Figure 30
Palaeogeography of Western Britain during the Lower and Middle Jurassic.

complex. Apart from a narrow opening to the east, through the Bristol Channel, the Celtic Sea depression is thought to have remained as a largely land-locked basin throughout Mesozoic times with only occasional and sporadic links to the north across the Irish-Welsh barrier into the north Irish Sea, and to the south across Cornubia into the Western Approaches Basin. This rapidly subsiding basin, and its northeastern arm through St George's Channel, first became flooded at the beginning of the Jurassic during the initial advance of the Liassic Sea across southern England. Marine conditions remained until the close of the Jurassic period, when renewed Late Kimmerian earth movements gradually elevated the area, causing the sea to retreat eastwards though the Bristol Channel leaving shallow water and eventually lagoonal and coastal plain conditions in this Celtic Sea depression. However, during the Middle Jurassic period there was a general shallowing of the sea over the southern half of Britain, with the earlier deep water clays and shales gradually giving way to shallow-water limestones and sandstones. Further uplift of the broad Welsh-Central Irish Island to the north caused the sea to retreat somewhat earlier from the shallow St George's Channel inlet and Upper Jurassic sediments were only sparsely developed here.

North of the Welsh landmass, the Liassic Sea spread westwards through a gap in the Pennine barrier flooding the northern Irish Sea and adjacent Lancashire-Cumberland coastline (see Figure 30). It also covered much of Northern Ireland and the entire western coast of Scotland and Inner Hebrides, eventually linking up with the Rockall Trough marine basin. During this period, most of the Scottish Highlands and Pennine Ridge formed a single elongate island.

Although well developed marine conditions existed both to the north of Scotland and over Northern Ireland to the south, a shallow water environment prevailed over the Inner Hebridean Basin for most of the Jurassic. During this period large quantities of coarse sediment were being introduced by rivers flowing from the large Greenland landmass to the northwest and dumped as coastal sandstones and deltaic features.

Towards the end of the Jurassic period the *Late Kimmerian* earth movements similarly began to affect northern England and Scotland, causing large scale uplift of the earlier depressions and a rapid southward retreat of the sea. Finally by end Jurassic and earliest Cretaceous times the sea had been excluded from the whole of the British Isles (with the exception of the Celtic Sea-Western Approaches depression and an associated basin in the extreme south) and there was a return once more to widespread continental conditions over Britain (see Figure 31). These continental conditions persisted throughout most of the Lower Creta-

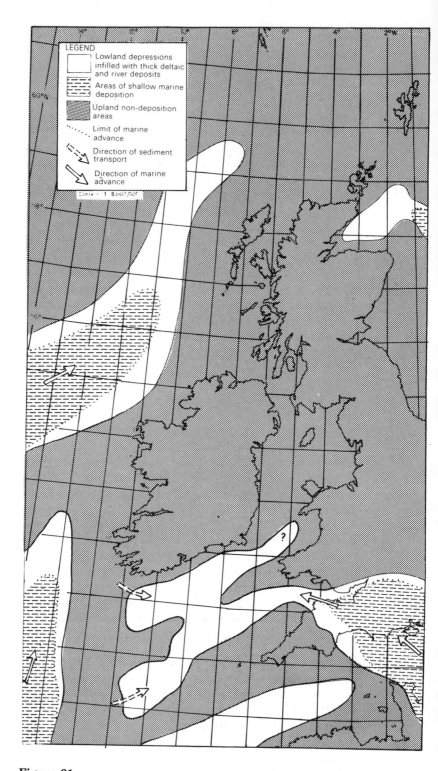

Figure 31
Palaeogeography of Western Britain during the Lower Cretaceous.

ceous and sedimentation during this period was characterised by large southeasterly flowing rivers which transported and then dumped sediments as deltaic fans in the restricted basin of the Celtic Sea and southern England.

Roughly concurrent with the northwestward extension of sea-floor spreading into the Labrador Sea, and the separation of the Greenland-European Plate from the North American Plate in Upper Cretaceous times, widespread regional subsidence affected the whole of Western Europe, and the British Isles themselves became rapidly submerged beneath a westward advancing sea. The sea first covered central and northern England, the Celtic Sea and Western Approaches before spreading westwards into the sea areas of the Porcupine Seabright and Rockall Trough. To the north, the margin of the Rockall Trough sea spread eastwards into Northern Ireland, western Scotland and the Inner Hebrides (see Figure 32). How much of the Welsh, Irish, Cornubian and Scottish mountainous tracts were submerged by this time is still uncertain, but it is believed that the central cores at least remained as positive island features while a thick chalk sequence was being unconformably laid down across the partially eroded surface of Middle and Upper Jurassic rocks forming the surrounding shelf region. In contrast, where the sea encroached on the coastline of these island features the chalk can be seen to form a considerably thinner deposit resting directly on an erosion surface of Lower Mesozoic or Palaeozoic rocks.

South of the shrunken Welsh-Irish island barrier, a thick accumulation of chalk was being deposited in the existing depressions of the Western Approaches, Celtic Sea and Cardigan Bay basins. However, away from these centres of deposition, the sea spread across their steep, often fault-bounded margins, to deposit a thin sequence of chalk across the submerged margins of the old Cornubian, south Welsh and southern Irish land areas. Following uplift during the Tertiary, much of these thin chalks have been removed by erosion and their earlier presence can now be inferred from a few scattered remnants.

North of the Welsh-Irish barrier, Upper Cretaceous chalk deposits are apparently absent over the north Irish Sea and most of the Inner Hebrides-West Scottish province, although indications of its former presence in at least the Inner Hebrides region are provided by scattered chalk fragments throughout the area. However, to the west and north, chalk was being deposited thickly in Northern Ireland and Rockall Trough and in the basins west of the Shetland and Orkney Islands respectively.

Towards the end of the Cretaceous the first pulses of the *Alpine* phase of mountain building, centred over southern Europe, began to affect Britain and Northern Europe causing large scale

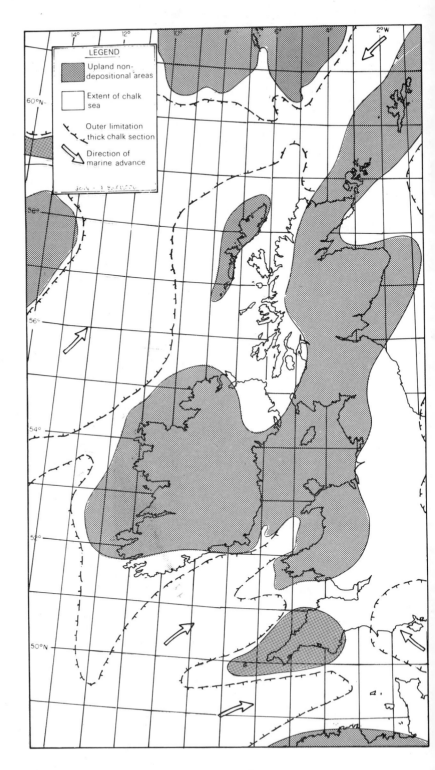

Figure 32
Palaeogeography of Western Britain during the Upper Cretaceous.

regional uplift and associated local downwarping. The Upper Cretaceous chalk sea slowly began to shrink away from the rising landmass of the British Isles, retreating eastwards and westwards towards the subsiding marginal depressions of the North Sea and Rockall Trough respectively. By Early Tertiary time, the palaeogeography began to assume a shape similar to that of the British Isles today. Land emerged to occupy the entire present-day land surface (apart from the extreme southeastern corner of England),including large tracts of the surrounding shallow continental shelf in the vicinity of the Shetlands and Orkneys, to the west of Scotland beneath the Sea of Hebrides and Minches, and further south over the northern Irish Sea, and to the west of Ireland over much of the Hatton, Rockall and Porcupine Banks.

Shortly after the commencement of the Tertiary period, the continental plate comprising Rockall and Western Europe finally separated from the Greenland Plate, and new oceanic crust extended northwards to underlie this northeastern part of the North Atlantic. The early phases of plate separation were accompanied by a period of marked crustal extension over Britain and the extrusion of large quantities of volcanic material at the earth's surface including widespread basaltic lava-flows across northwest Scotland, Northern Ireland, Rockall Platform and Greenland. In Mid-Tertiary times, as sea-floor spreading became established, the early extensional stresses ceased to dominate the British Isles and the structural pattern, to the south and east in particular, became gradually influenced by the marginal effects of the developing Alpine mountains to the south. Alpine movements over southern Europe and the Mediterranean region reached a climax towards the end of the Lower Tertiary period, throwing the thick marine sediments of the Tethys Sea into huge folds and overthrusts.

To the west of Britain, the Alpine movements played little part in altering the palaeogeographic pattern created·by the early stages of plate rifting, and the large areas of continental shelf which were emergent throughout most of the Tertiary only became submerged again in the quite recent Quaternary period, giving the indented coastline of the British Isles its present-day shape.

Selected Reading

WILLS L. J.   1952   Palaeogeographical Atlas
                                    Blackie and Son Ltd

ZIEGLER P. A.   1974   North Sea basin history in
                                    the Tectonic Framework
                                    Northwestern Britain

                                    Conference on Petroleum and
                                    The Continental Shelf of
                                    North West Europe (Institute
                                    of Petroleum: in press)

12 History of Oil and Gas Exploration

Before concentrating on hydrocarbon exploration in the areas west of Britain, the history of exploration in and around the British Isles will be briefly reviewed. Natural seepages of oil at the ground surface have been recognised for centuries in different part of Britain in rocks of various ages. Seepages have been recorded in places as far apart as Coalbrookdale, in Shropshire, and Lulworth Cove, in Dorset, although by far the most common local occurence of oil was associated with the collieries in the north of England. The volumes of oil involved in surface seeps is usually very small.

Natural gas was accidentally discovered whilst drilling a water well at Heathfield in Sussex in 1895, and until recently was used to light the local railway station. The demand for petroleum during the period of the First World War prompted the British Government to seek indigenous oil supplies and a number of wells were drilled after the war with limited small success. From that time up until the last five years, most of the exploratory drilling in the British Isles was carried out by British Petroleum. In the period 1936 to 1968 about 20 small fields were discovered with the relatively insignificant cumulative production of 15×10^6 barrels of crude oil, by comparison with the *daily* consumption of 2×10^6 barrels of oil in Britain at the present time. Many of these small onshore fields occur in the Carboniferous, and particularly the Coal Measure, rocks in the north of England. However, interesting fields also produce from Jurassic rocks in the Kimmeridge-Wareham area of Dorset. Although a number of the older fields are now depleted and no longer produce oil, the phenomenal success of the offshore North Sea area combined with the rising price and sarcity of oil is leading to a new phase of drilling activity onshore in Britain in which deeper rock horizons are likely to be tested.

The offshore search for oil was prompted by the discovery in 1959, near Groningen in north Holland, of vast gas reserves in sandstones of Permian age at a depth of almost 10,000ft. These sandstones are wind and river lain desert deposits and the source of the gas is almost certainly in the older Carboniferous Coal Measure beneath. The search for oil beneath the North Sea clearly required that the national ownership of the offshore areas be determined. This was achieved in 1964 at a United Nations Convention in Geneva, at which the North Sea was subdivided into seven sectors, each sector being allocated to the respective bordering nation.

Following the discovery of several large gas fields in the southern part of the North Sea on a latitude with the Groningen discovery, the exploration effort gradually moved northwards into the northern part of the North Sea, with the result that in the early 1970's a series of massive oil finds were discovered in British and Norwegian waters in marine rocks of Mesozoic and Tertiary age. The impetus given to the offshore search by these discoveries has now resulted in the gradual shift of interest to the virgin areas west of Britain.

The 1964 Geneva Convention which determined the position of median lines between the countries around the North Sea, did not settle the position of the median lines to the south and west of the British Isles, thus the location of the offshore median lines between the Republic of Ireland, Britain and France are still in dispute. In a sense, each new discovery makes it more difficult to come to an agreement, as the possible importance of small shifts in the position of the line are realised.

Under the provisions of the Geneva Convention of April 29th, 1958, a nation may exercise sovereignty over:

i   The territorial area of the sea-bed and submarine zone adjacent to the coast; also outside the boundary of the territorial area up to a depth of 200 metres (657ft). Beyond this limit, up to a point at which the depth of the superjacent waters acts as a technical limit to the exploitation of the natural resources of the said areas, and

ii   The sea-bed and subsoil of submarine areas adjacent to the coasts or islands.

In fact, many nations, including both Ireland and Britain, have designated deeper waters than 200m and in some cases granted licences in them. Britain has already granted all or part of 18 deep water licences in the northern North Sea and also 17 licences west of the Shetlands, while the Republic of Ireland has designated deep water areas in the region west of the Porcupine Bank. The United Nations is considering this problem together with the whole problem of the national ownership of the resources of the deeper oceanic regions. In the meantime, however, technology and the demand for petroleum is pushing out national offshore boundaries into progressively deeper water.

The position of the median lines and the extent of designated waters around each country is shown on Figure 33. As will be seen, only a narrow corridor is under dispute in the Celtic Sea between Ireland and England; while to the south of Britain the position of median line between France and Britain is far from

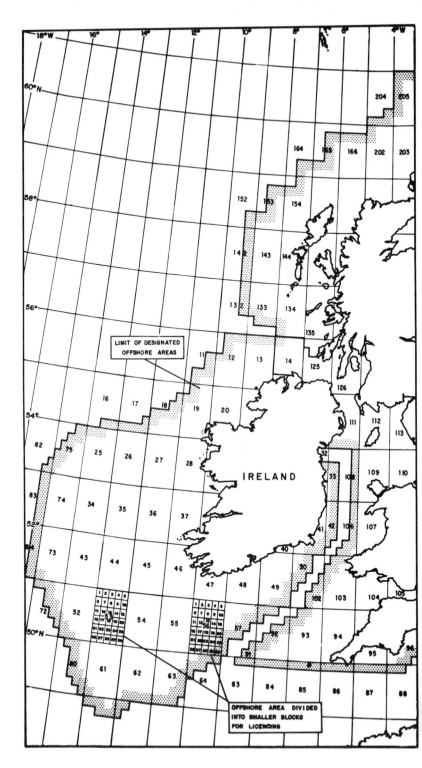

Figure 33
Licensing system of the offshore designated areas.

settled, chiefly because of the location of the Channel Islands close to the French coast. A line drawn southwards from British waters and around the Channel Islands would clearly exclude the French from sovereignty over much of the Channel.

The procedure for allowing and containing offshore exploration is similar throughout most of the countries of Western Europe. Initially an area of water is designated by the government of the country concerned, which effectively takes that water under the jurisdiction of the government for the purpose of exploration. It is normally possible at this stage for companies to acquire non-exclusive licences which allow for the exploration of, but not production of, petroleum. It is during this phase that wide-grid regional seismic data is usually obtained — either by exclusive one-company shooting, or by groups of companies acting together or by purchase of speculative seismic data from seismic contracting companies. Normally the following stage is for the government to offer and grant offshore blocks within the designated area. The size of these blocks in Irish waters is shown on Figure 33. A company then has exclusive rights on the block for a period usually of about six years. During this phase, the company would then acquire more detailed seismic coverage over any interesting subsurface structures and eventually drill a test well. In the event of a discovery, then it is possible for the company concerned to convert their exploration licence to some form of production licence which allows for the production of petroleum from that field over a period of years.

There are slight differences between the various Western European countries attitude towards licencing but the overall method is very similar. In general the main difference between the countries is in the terms asked of the companies in return for licences. In the past, these have normally been very modest — usually a promise to do a certain amount of work on the block. Since large amounts of petroleum have been found, however, this attitude is changing and in future a high government participation is to be expected in most areas. Normally the government would expect half of any discovery (government costs to be paid from production) in addition to any royalties or taxes. A total government share in the order of 75-80% can be anticipated.

We have seen that the first probings of the offshore areas west of Britain and Ireland were made by universities and government research teams. This work has continued and some aspects, such as bottom sampling, are not normally repeated by the oil industry. With awakening interest in the western areas, oil companies have begun to acquire high quality seismic data, normally too expensive to be shot by university teams. The first speculative seismic survey, making data generally available to the industry, was shot in 1970 in the area west of Shetlands. The speed with which the industry has moved since then may be judged by the fact that there are now more than 40,000 miles of data available for purchase in the areas west of the British Isles. In addition to these data there has been a considerable amount of seismic shooting on behalf of individual companies. The result is that even in the unlicenced areas there is a great deal of detailed knowledge now available to the oil industry. Thus at an early stage in exploration history it is possible for a company, and also a government, to take an overview of a whole region. This is in contrast to exploration in the past onshore in continental areas where exploration has tended to work gradually into a sedimentary basin, and where it has taken several decades for the overall basin architecture to become apparent. In turn, this has had an impact on the philosophy of exploration.

In fact, the first offshore exploration well in British waters was drilled in 1963 by BP (Lulworth Banks No 1) in the English Channel to a depth of 2,500ft. The well was 8 miles west of the small onshore oilfield at Kimmeridge which produces from Jurassic limestones and fractured shales. However, the Channel area is one of the world's busiest seaways and this will make further offshore drilling very hazardous. In addition, as mentioned above, there is the thorny problem of the, as yet, undefined median line between Britain and France.

In the West Shetland area the first speculative offshore seismic data was shot by Delta/Seiscom in 1970. Fifty-eight blocks were allocated in the area at the United Kingdom Fourth Round of licencing, to some 19 groups or companies. Esso drilled the first hole in this area in August 1972 which proved unsuccessful, and it is generally accepted that the well was located on a basement ridge covered by only 5,000—6,000ft of young sediments. Further wells followed during 1974 but as yet without commercial success.

The Republic of Ireland has had a quite different history of petroleum exploration. The onshore area has not had the same degree of exploration effort as onshore Britain. Onshore concessions have been held in the Republic only since 1960, first by Ambassador Oil Company and then by an Ambassador-Conoco-Marathon Consortium. Subsequently there followed a small unsuccessful exploration programme aimed at the limestones and sandstones of the Carboniferous.

After a first relinquishment of 50% in 1965, Marathon (the remaining member of the original group) severely curtailed exploration activity. From about 1965 they were negotiating with the Irish Government about the offshore areas. According to the original agreement, the concession holder was entitled to the entire continental shelf when this was designated by the government. However, Marathon Oil Company reduced their claim to the three tracts which comprise about 50% of the entire continental shelf. Part of the agreement was that Marathon would reduce their area under permit by 25% in 1970 and a further 25% in 1975. The Irish Government initially designated an area slightly larger than the block finally granted to Marathon. This was extended during 1970 to cover their entire continental shelf and the Porcupine Seabight below the 600' bathymetric contour, in order to be able to control exploration better. A further extension was added in early 1974 to carry the designated area into the deep water west of the Porcupine Bank.

Since acquiring exploration rights, Marathon has carried out a steady drilling programme in their acreage off the southern coast of the Republic. Subsequently in 1972 Esso farmed-in to the western part of the Marathon acreage with a commitment to drill a number of wells. Both companies have now drilled more than 13 wells, many of which have had rumoured hydrocarbon shows. One well has resulted in the discovery of the Kinsale Head gasfield, situated 30 miles south of Cork (see Figure 14) by Marathon, which has recently been declared commercial with reserves of about 1 trillion cubic feet of gas. It is probable that Lower Cretaceous sandstones form the pay zone in this field.

Gulf/National Coal Board, in partnership, drilled two wells in 1969 in the Irish Sea, to the east of the Isle of Man. This region forms the Manx-Furness Basin where thick Permo Triassic reservoir horizons overlie potential source rocks of Carboniferous age. The first was drilled to 7,000ft and the second to 10,400ft. Neither of the wells encountered commercial quantities of hydrocarbons, although it is believed that significant gas shows were found. A third well drilled in 1974 was announced as a gas discovery but no details were released.

If any single feature symbolises the offshore areas west of Britain, it is the precipitous islet of Rockall rising starkly from the stormy waters of the North Atlantic, 300 miles west of Scotland. The islet, only seventy feet high, was taken possession of in 1955 by a landing party from HMS Vidal. Since that time the British Admiralty have placed a navigational beacon on the top of the rock. By passing the Rockall Act 1972, the British Parliament incorporated Rockall into the county of Inverness (Scotland) and by so doing acquired the right to designate the areas around the islet. Designation was carried through in September 1974. Clearly the question of sovereignty in these oceanic areas west of Britain and the location of a median line between the Faroes, Rockall and Ireland must be decided. The government of the Republic of Ireland now disputes British ownership of Rockall on the grounds that Irish missionaries had known of and visited the island many centuries before. Certainly, Rockall epitomises the stormy condition west of the British Isles which the oil industry must combat in the next decade, and also the political uncertainties attached to offshore exploration.

Selected Reading

NAYLOR, D   1972   The hydrocarbon potential of Western Britain and Ireland. North Sea Conferences 1 and 2 IPC Industrial Press130-137.

WHITBREAD, DRW   1972   The hydrocarbon potential of Western Britain and Ireland. North Sea Conferences 1 and 2. IPC Industrial Press 81-94.

Geology of the North-West European Continental Shelf
Volume 1

13 Appendix

13.1   Basic Stratigraphical and Structural Features

Sedimentary strata: A vertical accumulation of sedimentary rocks displaying a layered structure.

Basement: Considered here as synonomous with an agneous or metamorphosed rock of Precambrian or Palaeozoic age which is below the limit of economic importance for hydrocarbon exploration.

Pinch out: Term applied to wedge-like feature where the total thickness of a sedimentary stratum is gradually reduced to nothing. Formed as an original feature where sediments are deposited against a raised margin, or as a post-depositional feature resulting from partial erosion of the stratum.

Onlapping Sediments: Feature occurs where progressively younger members of a younger stratum rest on a dipping older surface.

Reef: A dome-like mass of organic skeletal material consisting chiefly of organisms which have grown in situ over a period of time. The geological conditions permitting reef growth are limiting, and their occurence is therefore a good indicator of the geological environment existing at that time.

Unconformity: A plane which separates older rock from younger rock and indicates a break or interruption in a phase of continuous sedimentation. Such a plane may appear as either an irregular surface cross-cutting the underlying older rocks, or may be closely parallel or parallel to the overlying strata, depending upon its mode of origin, and as such may represent either an erosional surface, a surface of non-deposition or a combination of both.

Anticline: An arch-shaped flexure or upfold of the strata.

Syncline: A basin-shaped flexure or downfold of the strata.

Horst: A plateau-like block of crust forced up between two roughly parallel faults.

Graben: A block of crust which subsides or is let down between two roughly parallel faults. Also termed **Rift Valley.**

Igneous intrusion: Large mass of molten material, orignating in the lower crust, intruded into the upper crustal strata. The mass is exposed as an approximately circular crystalline body following erosion of the overlying sediments. Heat dissipated from this intruded body often causes alteration of the surrounding sediments as it cools and solidifies.

Dyke: A sheet-like body of igneous material cross-cutting the normal stratigraphic layering of the crustal material.

Sill: A sheet-like body of igneous rock arranged parallel to the layering of the upper crustal strata. Both dykes and sills frequently occur in association with larger igneous intrusions.

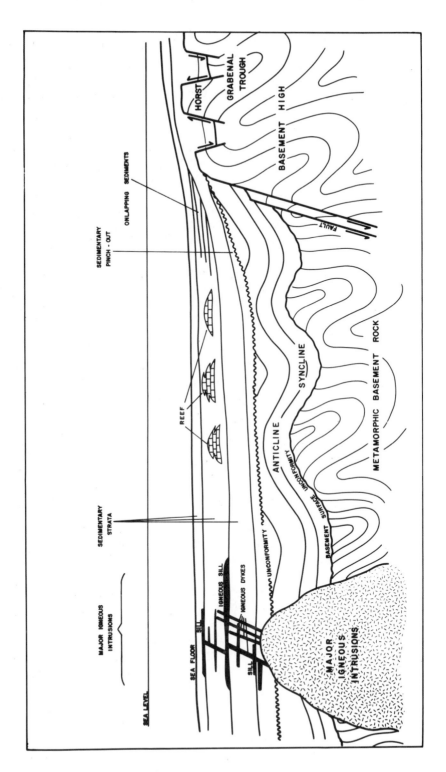

Figure 34
Basic stratigraphical and structural features.

13.2   Geological Control for Oil and Gas Entrapment

Four basic geological conditions are required for the accumulation and retention of commercial quantities of hydrocarbons in the form of either natural gas or oil by itself, or in combination with a gas cap overlying the oil. The geological requirements are as follows:

1   The presence of an organic rich *source rock* such as coal, deltaic, marine or lagoonal sediments, from which hydrocarbons can be generated.

2   A suitable *reservoir rock* such as sandstone, either directly above or adjacent to, and in more unusual cases beneath, the source rock, formed by a both porous and permeable sedimentary layer into which the hydrocarbons can migrate and be stored.

3   The presence of a *cap rock* or impermeable strata, such as clay, shale or mudstone, which acts as a seal to prevent the hydrocarbons escaping to the surface.

4   A *trapping mechanism* in which hydrocarbons collect and are prevented from either lateral or vertical escape to the surface.

Traps fall into two categories, which may be classified as either stratigraphical or structural according to their mode of origin.

Stratigraphical traps are formed at the initial time of sediment deposition where sediments were laid down across an uneven erosional surface, against an ancient shoreline, or even in ancient reefs, where the reef is surrounded by impermeable sediments.

Structural traps are formed during post-depositional times by both large-scale and local deformation of the sedimentary strata. These include:

Unconformity traps which occur where impermeable cap rock strata rests directly on the eroded surface of a dipping source, reservoir, cap rock sequence, trapping hydrocarbons beneath the unconformity.

Fold traps which are caused by widespread flexuring of the strata into anticlinal and synclinal folds trapping hydrocarbons in the anticlinal arches.

Fault traps which are formed where rocks have broken along a line of weakness shifting the strata on one side with respect to the other. Hydrocarbons are trapped where an impermeable rock is aligned against a reservoir rock and no escape is possible along the fault line.

Salt-domes and *salt-pillow traps* which are a common form of locally generated trapping mechanism where a reasonably thick salt layer exists as a part of the stratigraphical sequence beneath the source rock. Salt forms a relatively light density rock in comparison with most other sedimentary rocks and when confined at depth by the gradual increase in weight of the overlying sediments, an unstable gravity situation is created. Under high

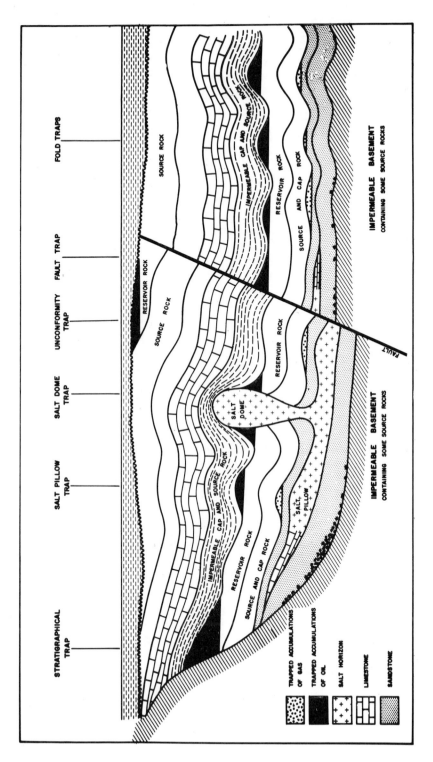

Figure 35
Geological controls for oil and gas entrapment.

149

pressure the salt will behave plastically, pushing up (salt pillows) and piercing through (piercement salt-domes) the overlying sediments in an attempt to reach gravity equilibrium, and in doing so, as an impermeable material, forming hydrocarbon traps as illustrated on the opposite page.

13.3   Seismic Exploration Methods

Seismic exploration provides a method of observing the thickness and structure of the individual rock strata forming the earth's crust. This method is based on the different reflecting and absorbing properties of the various rock types with respect to a shock wave generated at or near the earth's surface. A shock wave from a point source passes downwards into the underlying sediments, continuing at roughly the same velocity within any one sediment type until it encounters an interface, such as the boundary between two differing rock types. At this boundary, part of the shock wave is reflected back to the surface where it is picked up and recorded by a system of *geophones* (signal receivers), and part is refracted down into the underlying strata. Of this refracted part, a similar process of wave separation occurs at the next boundary down, with a somewhat weaker reflected part being returned to the surface. The speed at which shock waves travel through any rock is dependent upon the *elasticity* of that rock; this tends to be low in loose and uncompacted sediments, increasing with depth as the sediments become harder and more compacted, and higher in both igneous rocks such as basalts and granites, and metamorphosed rocks such as schists and slates. The passage of a shock wave through rock follows the normal physical laws of reflection and refraction, and the success of producing a good *seismic picture* is dependent upon an increase in the speed at which waves will pass through rock, with increased depth. With a prior knowledge, from refraction data, of the velocity of specific rock formations, the measurement of both the time taken for the wave to be reflected back to the surface and their strength, can be used to build up a seismic picture from which the depth and nature of the rock layers can be interpreted.

As an exploration method this can be adapted for both use on land and offshore. Offshore operations, which are of chief concern to the oil industry, are carried out from a boat equipped with an energy source for the generation of shock waves, and towing a string of geophones suspended on floats or strung out on a buoyant cable up to 4,800ft long. This cable is floated some 40ft beneath the water surface in order to minimise the interference of unwanted signals such as surface wave or boat noise. Explosives are now rarely used as shock wave generators, and the energy sources now adopted for deep water operations fall into two categories: the air gun source, used where deep penetration of the underlying strata is required, generates a seismic pulse by suddenly releasing highly compressed air into the water; while both the *sparker* and *boomer* sources, generated by an electrical-

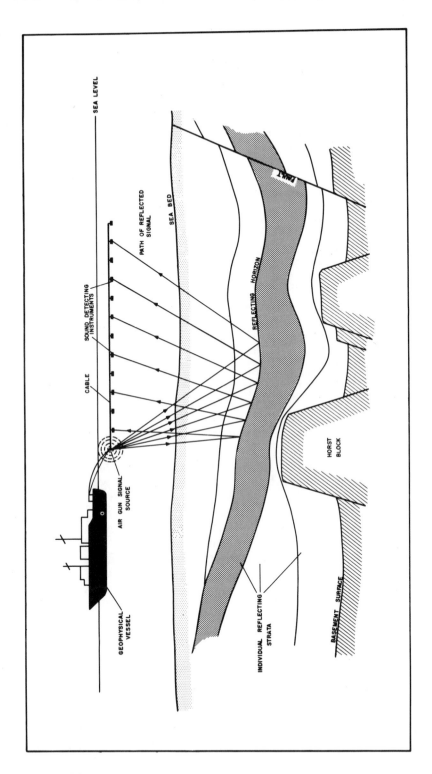

Figure 36
Diagram illustrating the method of obtaining offshore seismic profiles.

arc discharge device and an electromechanical device respectively, are used only where shallow rock penetrations are required. All signals from the reflected waves are measured as a time period in seconds between the signal being released, hitting the reflective boundary and being returned to the surface to be picked up by the geophones. These are fed back into a computer on board the boat, where they are processed to produce a geophysical picture in seconds based on the two way travel time of the underlying rock strata which takes the form of a *wiggly-line* paper record.

13.4   Drilling a well

Although initially exploration surveys can provide a good indication of the most likely location of a hydrocarbon-bearing structure, the only direct means of testing these prospects is by drilling a well. **Offshore,** and in particular around the shores of the British Isles there is a considerable variation in the depth to the continental shelf. This has led to the use of the three types of drilling rigs illustrated on the opposite page, each being suitable for working under different water depths and drilling conditions. *Jack-up rigs* are used in shallow water areas, where the water depths do not exceed 300ft and the sea-bed is firm. This type of rig is floated into position and the legs lowered to stand firmly on the sea-floor, with the drilling platform being jacked clear of high-water level. Where water depths are deeper or the sea-bed is unstable, a mobile floating type of rig, the *semi-submersible rig,* is employed. This is usually capable of working in water depths of up to 600ft and comprises a drilling platform supported on a partially submerged buoyant framework. The design of the semi-submersible provides a relatively stable platform under rough open-marine conditions which are subject to strong wave motions. More recent designs now enable some of these rigs to drill in water depths of over 1,500ft. *Drilling ships* are capable of operating in all areas, and in particular where water depths exceed 600ft.

An exploratory well basically comprises a hole between seven and thirty inches in diameter drilled into the upper crustal sediments to a depth of up to 30,000ft. The location of the hole is sited to intersect any prospective structural features of stratum picked out on the initial geophysical survey. The uppermost part of the hole is protected by a series of steel pipes or *casing* cemented to the rock wall of the hole. The walls of the greater length of the hole are unprotected, so that instruments inserted down the hole can establish the various physical properties of the rocks drilled through. The hole itself is drilled using a rotating steel-toothed or diamond-studded *bit* attached to the lower end of a heavy steel shaft (the *drill string*). Throughout the drilling programme a high

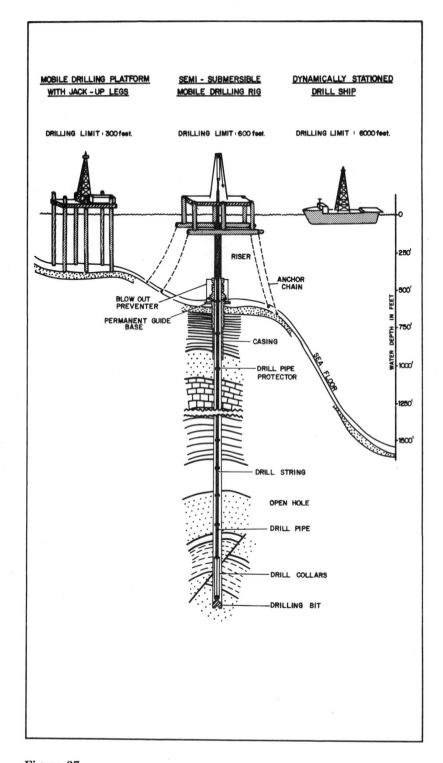

Figure 37

Diagram illustrating the three main types of offshore drilling rigs and a section through a typical exploratory well.

density mud is circulated through the hole, acting partly as a lubricant, a coolant, and to transport the newly excavated fragments from the bottom of the hole to the surface, and partly to prevent any inward collapse of the lower unprotected well walls, On the sea-bed itself the *blow-out preventor* consists of a complex array of pipes and valves installed to prevent any sudden seepage of oil or gas onto the sea-floor in the event of encountering a reservoir containing hydrocarbons at high pressures.

14 Glossary

A short glossary such as this can be neither systematic nor complete. The entries listed below should be taken as a rough guide to meanings of geological terms used in this book and not as formal definitions.

Alluvial　　　　　　　　Continentally eroded material which is tranported by a river and deposited at points along the river margin.

Anticline　　　　　　　　See Figure 34, and accompanying explanation.

Argillaceous　　　　　　Group term applied to fine-grained sedimentary rocks comprising clays, shales, mudstones, siltstones and marls.

Basalt　　　　　　　　　A fine-grained, dark (often black), igneous rock, forming lava flows and thin dykes and sills. Represents rapidly cooled material from the lower crust.

Basement　　　　　　　　See Figure 34, and accompanying explanation.

Breccia　　　　　　　　　Sedimentary accumulation of angular fragments of broken rock of very variable size.

Chalk　　　　　　　　　　Very fine-grained, pure white limestone.

Chert　　　　　　　　　　Nodules or bands or microscopic silica (SiO^2) crystals which may be of organic or inorganic origin. Flint is a variety of chert.

Calcareous　　　　　　　Containing calcium carbonate; usually applied to sedimentary rocks such as chalk or limestone.

Conglomerate　　　　　　Sedimentary accumulation of rounded or semi-rounded fragments of rock, implying rather more transport than *breccias*.

Continental shelf　　　That part of the sea-floor adjoining a land mass over which the water depth rarely exceeds 600ft. A continental shelf is considered to be a locally submerged portion of the continental mass, the outer margin of which is regarded as the boundary between continental crust and oceanic crust.

Crust of the Earth	The surface portion of the earth lying above the *Mantle*. The crust is divided into two shells: a lower continuous layer — the sima (oceanic type crust), and an upper discontinuous layer — the *Sial* (continental type crust which is confined to the continental masses only). The total thickness of the composite crust reaches a maximum beneath a mountain chain and a minimum beneath the oceans.
Denudation	The total process of weathering, erosion and sediment transportation causing the wearing away and eventual lowering of the land surface.
Dolomite	Limestone containing more than 15% magnesium carbonate.
Dyke	See Figure 34, and accompanying explanation.
Evaporite	The remains of a solution after most of the solvent (usually water) has evaporated e.g. salt
Facies	The collective features which characterise a sediment deposited in a particular sedimentary environment.
Limestone	A sedimentary rock consisting essentially of organically or chemically derived carbonate minerals.
Lithology	A term applied to sediments with reference to their general characteristics.
Mantle	The position of the mobile earth's interior which lies beneath the crust.
Marl	A calcareous mudstone.
Metamorphism	The process by which rocks of the earth's crust are altered through the agencies of heat, pressure and chemically active fluids.
Palaeogeography	A reconstruction of the relative positions of land and water at a particular period in geological history.
Permeability	The ability of a rock stratum to allow water or any other liquid to pass freely through from one surface to another or along its length.

Porosity	The occurence of voids or cavities between the mineral grains making up the internal structure of a rock, which give it an ability to retain fluids or gases. Porosity need not necessarily be accompanied by permeability.
Reef	See Figure 34, and accompanying explanation.
Schist	A regionally metamorphosed rock characterised by the thermal alteration of the original minerals and their recrystallisation into a roughly parallel direction.
Shield	A major structural unit of the continental crust, consisting of an extensive mass of Precambrian rocks which have remained unaffected by later earth movements.
Sill	See Figure 34, and accompanying explanation.
Slate	Regionally metamorphosed *argillaceous* rocks in which strong pressures have re-orientated the original minerals into a single direction.
Strata	See Figure 34, and accompanying explanation.
Syncline	See Figure 34, and accompanying explanation.
Tectonic	Structurally derived by earth movements and large scale deformations.
Thrust	A near horizontal plane of dislocation or fault.
Unconformity	See Figure 34, and accompanying explanation.

15 Index